PHYSICS RESEARCH AND TECHNOLOGY

FROM MAXWELL'S EQUATIONS TO FREE AND GUIDED ELECTROMAGNETIC WAVES

AN INTRODUCTION FOR FIRST-YEAR UNDERGRADUATES

PHYSICS RESEARCH AND TECHNOLOGY

Additional books in this series can be found on Nova's website
under the Series tab.

Additional e-books in this series can be found on Nova's website
under the e-book tab.

PHYSICS RESEARCH AND TECHNOLOGY

FROM MAXWELL'S EQUATIONS TO FREE AND GUIDED ELECTROMAGNETIC WAVES

AN INTRODUCTION FOR FIRST-YEAR UNDERGRADUATES

MANUEL QUESADA-PÉREZ

AND

JOSÉ ALBERTO MAROTO-CENTENO

New York

Library of Congress Cataloging-in-Publication Data

ISBN: 978-1-63117-453-7

Published by Nova Science Publishers, Inc. † New York

CONTENTS

PREFACE

Let's imagine how our everyday life would be without telephone, radio, television, mobile phones, internet… How would we communicate with each other at a distance? In the information age, it is quite difficult to imagine a world without telecommunications. Maxwell's equations and the discovery of electromagnetic waves changed the world making this technological revolution possible. It's thanks to them that we understand what electromagnetic waves are and how they are generated, propagated and detected. Many students of physics and engineering must face Maxwell's equations sooner or later during their degree. In some cases, this moment can even be frightening. But these equations eventually change their perception of nature when they are really understood.

We can find Maxwell's equations in many textbooks of physics for first-year undergraduates but, in most of them, these equations are written only in integral form. However, the power and elegance of Maxwell's equations can be completely appreciated when they are expressed in differential form. Moreover, this form is extremely useful dealing with some issues, such as the propagation of electromagnetic waves. In general, only textbooks of advanced electromagnetism courses offer Maxwell's equations in differential form and discuss its applications to electromagnetic waves. However, the conceptual and mathematical level of advanced courses is too high for first-year students.

Thus there is a gap between the introductory level in which Maxwell's equations and electromagnetic waves are addressed in books for first-year undergraduates and the level of advanced textbooks, which involves sophisticated mathematical tools (differential equations, vector operators, complex analysis). This book will help students to bridge the gap between elementary and advanced treatments on Maxwell's equation and electromagnetic waves.

Our work is mainly aimed at undergraduates that must face the differential form of Maxwell's equations and its application to electromagnetic waves for the first time. On the one hand, this book will help them to unravel the secrets of an elegant and sophisticated notation, which can be indecipherable and discouraging for beginners. On the other hand, we offer illustrative examples of how this new mathematical expressions allow addressing problems such as the propagation of electromagnetic waves in vacuum and confined media. And we intend to achieve these goals with the mathematical level of first-year students. Obviously, the book can also be extremely useful for professors teaching these topics to such students. We encourage them to read the book. They could find valuable tools for this task.

The first chapter is devoted to Maxwell's equations in differential form. The main goal is that students facing this topic for the first time get to know a mathematical form that is not usually included in introductory textbooks of physics. This chapter also presents the notation based on vector analysis, which leads to highly elegant and compact expressions. The first chapter also shows how the differential form is deduced from the integral form. Students are assumed to know the integral form, which is generally treated in introductory physics courses. However, it should be stressed that elementary differential calculus is preferred rather than more advanced mathematical concepts, which are not so widely known. For instance, students will not require previous knowledge on advanced vector calculus, theory of functions of complex variable or Taylor expansions. For a similar reason, vectors and vector operators are usually expressed and employed in three-dimensional Cartesian coordinate systems. In any case, our objective is that students understand how Maxwell's equations in differential form work analyzing illustrative examples. The resolution of such equations goes beyond the scope of this book.

Although this work is mostly focused on Maxwell's equations in empty space, the first chapter also includes two appendixes in which the changes required for their application to matter are outlined. The motivation is twofold. On the one hand, new useful concepts (such as the electric displacement) are briefly introduced here. On the other hand, some of these concepts offer additional examples of use of the divergence and the curl.

The propagation of plane electromagnetic waves is discussed in the second chapter. In this case the main aim is to understand how the differential wave equation can be deduced from Maxwell's equations in differential form, without requiring complex identities of vector calculus. In this way, first-year undergraduates can reach a better comprehension of Maxwell's milestone: He showed the electromagnetic nature of light for the first time.

The third chapter deals with the propagation of electromagnetic waves between two conducting planes and along a rectangular waveguide. These are just two particular cases of guided waves but, from a pedagogical viewpoint, they offer important advantages:

i)   they allow visualizing the mechanism of propagation in confined media;

ii) these waveguides can be analyzed with elementary mathematical tools;

iii) the peculiarities of the propagation in confined media (as compared to free space) can be easily deduced and understood. Guided waves in other geometries require advanced mathematical treatments, which sometimes hide the physical understanding of phenomena.

In addition, each chapter ends with a short text addressing some concepts of the chapter from a historical perspective. These historical notes can be really motivating and enjoyable for those students who want to go beyond purely academic aspects.

ABOUT THE AUTHORS

Manuel Quesada-Pérez and José Alberto Maroto-Centeno are Professors of the Department of Physics in the Polytechnic School of Linares, University of Jaén, Spain. They hold a Ph.D. degree in physics from the University of Granada. Their research has focused on a number of issues about colloid science and soft matter, including optical techniques such as light scattering. In the last five years, Manuel Quesada-Pérez and José Alberto Maroto-Centeno have published more than 30 articles related with these research activities. In addition, they have published several works on teaching in prestigious international journals and books. Their lectures on electromagnetism for telecom undergraduates have inspired this work.

We shall find that it is the peculiar function of physical science to lead us . . . to the confines of the incomprehensible

J. C. Maxwell

Chapter 1

Maxwell's Equations in Differential Form

1.1. Introduction

Maxwell's equations in integral form relate electric and magnetic fields that exist in regions of arbitrary dimensions to their sources. In the case of Gauss's law, the flux of the electric field is evaluated over relatively large surfaces, which can be located far away from the source of such a field (electric charge). In the case of Faraday's law or Ampère-Maxwell's law, line integrals are calculated along arbitrarily long closed paths. The integral forms of Maxwell's equations can be particularly useful when we want to calculate electric or magnetic fields originated by symmetric charge or current distributions. The electric field created by a uniformly charged sphere or the magnetic field generated by an infinite rectilinear wire carrying a steady current are classical examples of these applications. Be sure that you know the integral form of Maxwell's equations and some of their typical uses before reading this book. The integral formalism is our starting point and nowadays there are many excellent textbooks addressing these issues at introductory level. However, the integral form of Maxwell's equations is not the most appropriate one for some situations involving variable electromagnetic fields (such as the propagation of electromagnetic waves). In these cases, the differential approach is much more efficient. In this mathematical form, Maxwell's equations also relate electric and magnetic fields to their sources (electric charges, currents and variable fields), but this relationship is restricted to the immediate neighborhood of a given point. Recall that derivatives are used in mathematics to study local properties of functions. Before tackling the study of electromagnetic waves, we must therefore transform the mathematical aspect of Maxwell's equations. This is just the main goal of this chapter. After this task, the integrals over surfaces and lines appearing in such equations will be replaced by differential operators with a very different appearance. However, the essential physical meaning of these famous electromagnetic equations (the relation between fields and sources) will remain unaltered.

1.2. Gauss's Laws in Differential Form

1.2.1. Divergence of the Electric Field

The main goal of this section is transforming the integral form of Gauss's law:

$$\Phi_E = \oint_S \vec{E} \cdot d\vec{A} = \frac{Q_S}{\varepsilon_0} \tag{1.1}$$

This law is expressed in terms of the flux of the electric field (Φ_E) through a closed surface (S), which is the surface integral appearing in Equation 1.1. The right side of this equation is the total amount of charge enclosed by S divided by the permittivity of free space. In principle, the closed surface can be as large as desired. However, we are interested in a given point of space, P, whose Cartesian coordinates are (x, y, z), and its immediate neighborhood (see Figure 1.1).

Thus we must compute the flux through a closed surface with quite small dimensions. Consider a rectangular prism centered at P, whose edges parallel to x-, y- and z-axis have lengths Δx, Δy and Δz, respectively, as shown in Figure 1.1. The flux through this prism can be split into the sum of the fluxes through each of its six faces, to which we will refer as 1, 2, .., 6:

$$\Phi_E = \Phi_{E1} + \Phi_{E2} + ... + \Phi_{E6} \tag{1.2}$$

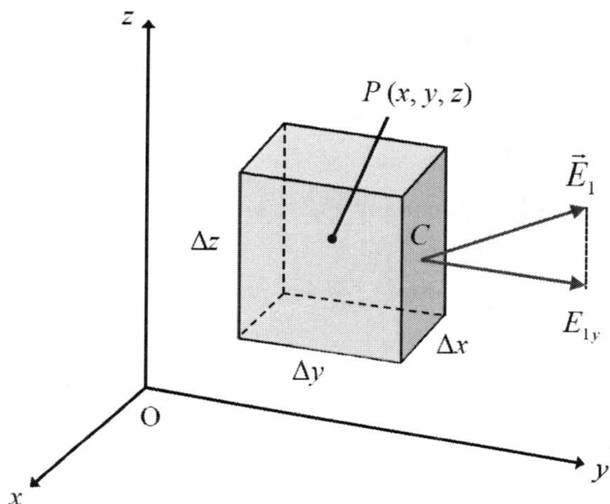

Figure 1.1. Rectangular prism with small dimensions centered at *P*. The electric field at one of its faces and the component perpendicular to this face are also shown.

Consider first the flux through face 1 (Φ_{E1}), which is one of the two faces parallel to the *xz*-plane. If the dimensions of this face are small enough, the flux can be estimated approximating the electric field at any point of this face by the electric field at its center (*C*), which will be represented by $\vec{E}_1$. This estimation can be written as:

$$\Phi_{E1} = \int_{face1} \vec{E} \cdot d\vec{A} \cong \vec{E}_1 \cdot \vec{A}_1 \tag{1.3}$$

where $\vec{A}_1 = \Delta x \Delta z \hat{j}$ is the vector area associated to face 1. Inserting this expression into Equation 1.3 gives:

$$\Phi_{E1} \cong \vec{E}_1 \cdot \hat{j} \Delta x \Delta z \tag{1.4}$$

Recall that $\vec{E}_1 \cdot \hat{j}$ is the *y*-component of the electric field at the center of face 1 (E_{1y}) and consider this component a function of position. Since the center of face 1 has the Cartesian coordinates (x, $y+\Delta y/2$, z), E_{1y} can also be expressed as $E_y(x, y + \Delta y/2, z)$. Thus, the flux through this face would be given by:

$$\Phi_{E1} \cong E_y(x, y + \Delta y/2, z)\Delta x \Delta z \qquad (1.5)$$

Consider now the flux through the face opposite to face 1, which will be referred to as face 2. Its Cartesian coordinates are $(x, y-\Delta y/2, z)$. Taking into account that $\vec{A}_2 = -\Delta x \Delta z \hat{j}$, it can be easily shown that:

$$\Phi_{E2} \cong -E_y(x, y - \Delta y/2, z)\Delta x \Delta z \qquad (1.6)$$

Thus the net flux through these two opposite faces (both perpendicular to the y-axis) can be written (after applying the distributive property) as:

$$\Phi_{E1} + \Phi_{E2} \cong \left(E_y(x, y + \tfrac{1}{2}\Delta y, z) - E_y(x, y - \tfrac{1}{2}\Delta y, z)\right)\Delta x \Delta z \qquad (1.7)$$

Doing likewise for faces 3 and 4 (perpendicular to the x-axis) and 5 and 6 (perpendicular to the z-axis) yields:

$$\Phi_{E3} + \Phi_{E4} \cong \left(E_x(x + \tfrac{1}{2}\Delta x, y, z) - E_x(x - \tfrac{1}{2}\Delta x, y, z)\right)\Delta y \Delta z \qquad (1.8)$$

$$\Phi_{E5} + \Phi_{E6} \cong \left(E_z(x, y, z + \tfrac{1}{2}\Delta z) - E_z(x, y, z - \tfrac{1}{2}\Delta z)\right)\Delta x \Delta y \qquad (1.9)$$

The flux through the prism is finally obtained adding the fluxes through the six faces:

$$\begin{aligned}
\Phi_E \cong &\left(E_y(x, y + \tfrac{1}{2}\Delta y, z) - E_y(x, y - \tfrac{1}{2}\Delta y, z)\right)\Delta x \Delta z + \\
&+\left(E_x(x + \tfrac{1}{2}\Delta x, y, z) - E_x(x - \tfrac{1}{2}\Delta x, y, z)\right)\Delta y \Delta z + \\
&+\left(E_z(x, y, z + \tfrac{1}{2}\Delta z) - E_z(x, y, z - \tfrac{1}{2}\Delta z)\right)\Delta x \Delta y
\end{aligned} \qquad (1.10)$$

As can be seen, this flux vanishes when the prism dimensions tend to zero. For this reason, the divergence is defined as the flux *per unit volume* through an infinitesimal closed surface surrounding the point of interest:

$$div\vec{E} \equiv \lim_{V \to 0} \frac{\Phi_E}{V} \qquad (1.11)$$

The divergence is therefore a scalar property associated to *each point of space*. Its physical meaning will be elucidated later with the help of a few examples. The flux per unit volume is obtained dividing the net flux by the volume of the prism $V = \Delta x \Delta y \Delta z$:

$$\frac{\Phi_E}{V} \cong \left(\frac{E_x(x+\frac{1}{2}\Delta x, y, z) - E_x(x-\frac{1}{2}\Delta x, y, z)}{\Delta x} \right) +$$
$$+ \left(\frac{E_y(x, y+\frac{1}{2}\Delta y, z) - E_y(x, y-\frac{1}{2}\Delta y, z)}{\Delta y} \right) + \tag{1.12}$$
$$+ \left(\frac{E_z(x, y, z+\frac{1}{2}\Delta z) - E_z(x, y, z-\frac{1}{2}\Delta z)}{\Delta z} \right)$$

Taking the limits $\Delta x \to 0, \Delta y \to 0, \Delta z \to 0$ yields:

$$div\vec{E} = \lim_{V \to 0} \frac{\Phi_E}{V}$$
$$\cong \lim_{\Delta x \to 0} \left(\frac{E_x(x+\frac{1}{2}\Delta x, y, z) - E_x(x-\frac{1}{2}\Delta x, y, z)}{\Delta x} \right) +$$
$$+ \lim_{\Delta y \to 0} \left(\frac{E_y(x, y+\frac{1}{2}\Delta y, z) - E_y(x, y-\frac{1}{2}\Delta y, z)}{\Delta y} \right) + \tag{1.13}$$
$$+ \lim_{\Delta z \to 0} \left(\frac{E_z(x, y, z+\frac{1}{2}\Delta z) - E_z(x, y, z-\frac{1}{2}\Delta z)}{\Delta z} \right)$$

At this point, you should recall the definition of partial derivative with respect to a given variable (say x):

$$\frac{\partial f}{\partial x} \equiv \lim_{\Delta x \to 0} \frac{f(x+\frac{1}{2}\Delta x, y, z) - f(x-\frac{1}{2}\Delta x, y, z)}{\Delta x} \tag{1.14}$$

From Equation 1.13 and the definition of partial derivative, it can be easily inferred that:

$$div\vec{E} = \left(\frac{\partial E_x}{\partial x} + \frac{\partial E_y}{\partial y} + \frac{\partial E_z}{\partial z} \right) \tag{1.15}$$

This expression is extremely useful to calculate the divergence when the Cartesian components of the electric field are known as function of position. The previous discussion not only provides this way of calculating the divergence taking partial derivatives, but also offers a geometrical interpretation of such derivatives: According to Equations 1.10, 1.11 and 1.13, $\partial E_x / \partial x$, $\partial E_y / \partial y$ and $\partial E_z / \partial z$ are the net fluxes per unit of volume in the x-, y- and z-direction, respectively (this means through infinitesimal surfaces perpendicular to the x-, y- and z-axis).

On the other hand, the divergence of the electric field can also be computed from Gauss's law:

$$div\vec{E} = \lim_{V \to 0} \frac{\Phi_E}{V} = \lim_{V \to 0} \frac{Q}{V\varepsilon_0} = \frac{1}{\varepsilon_0} \lim_{V \to 0} \frac{Q}{V} = \frac{\rho}{\varepsilon_0} \tag{1.16}$$

where ρ is the local charge density, defined as $\rho \equiv \lim_{V \to 0} Q/V$. Equation 1.16 is Gauss's law in differential form. If the divergence is written in terms of the Cartesian components of the electric field, the differential form of Gauss's law can be expressed as:

$$\frac{\partial E_x}{\partial x} + \frac{\partial E_y}{\partial y} + \frac{\partial E_z}{\partial z} = \frac{\rho}{\varepsilon_0} \tag{1.17}$$

In any case, this law relates the *local* behavior of the electric field at a given point of space to the charge density at such a point, that is, the law relates the field to the source. The mathematical aspect has considerably changed but the essential physical meaning remains the same.

1.2.2. A Compact Differential Form for Gauss's Law

The sum of the three derivatives appearing in the previous equation can be written in a more compact symbolic form with the help of the vector operator called *nabla* (and also *del*), represented by an inverted uppercase delta, $\vec{\nabla}$. This operator can be considered as a 'vector' whose three Cartesian components are three operators instead of three scalar quantities:

$$\left(\hat{i}\frac{\partial}{\partial x} + \hat{j}\frac{\partial}{\partial y} + \hat{k}\frac{\partial}{\partial z} \right) \equiv \vec{\nabla} \tag{1.18}$$

According to this definition, the sum of three derivatives can be seen as a dot (scalar) product of nabla and the electric field:

$$\frac{\partial E_x}{\partial x} + \frac{\partial E_y}{\partial y} + \frac{\partial E_z}{\partial z} =$$
$$= \left(\hat{i}\frac{\partial}{\partial x} + \hat{j}\frac{\partial}{\partial y} + \hat{k}\frac{\partial}{\partial z} \right) \cdot \left(\hat{i}E_x + \hat{j}E_y + \hat{k}E_z \right) = \vec{\nabla} \cdot \vec{E} \tag{1.19}$$

Within this notation, Gauss's law is finally expressed as:

$$\vec{\nabla} \cdot \vec{E} = \frac{\rho}{\varepsilon_0} \tag{1.20}$$

1.2.3. Gauss's Law for Magnetism

Magnetic field lines do not originate and terminate on anything. They form closed loops or straight lines without origin and end. Thus, for every magnetic field line entering a closed surface there must be a magnetic field line leaving it. In other words, the net number of magnetic field lines passing through a closed surface (i.e., the magnetic flux) is identically zero:

$$\Phi_B = \oint_S \vec{B} \cdot d\vec{A} = 0 \tag{1.21}$$

This result is also intimately related to the absence of isolated magnetic poles. As you can guess, the integral over a closed surface is also replaced by the divergence when Gauss's law for magnetism is transformed into a differential form:

$$\vec{\nabla} \cdot \vec{B} = 0 \tag{1.22}$$

1.2.4. Example 1: Point Charge

Although the electric field of a point charge is widely-known, it is quite instructive to verify that this field satisfies Gauss's law in differential form. This task can show you how the different concepts involved in this law work. Without loss of generality, let us consider the case of a point charge situated at the origin of the coordinate system. In this case, the electric field originated at any point of space (with the exception of point where the charge is located) is:

$$\vec{E} = K\frac{q}{r^2}\hat{r} = K\frac{q}{r^3}\vec{r} \tag{1.23}$$

where $\vec{r} = x\hat{i} + y\hat{j} + z\hat{k}$ is the position vector of the point where we are evaluating the electric field and $\hat{r} = \vec{r}/r$ is a unit vector pointing in the same direction. The goal is to show that this field verifies Gauss's law in differential form (Equation 1.17) at any point (except the origin). According to the mentioned equation, you must calculate partial derivatives of the Cartesian components of the electric field. Thus you should first identify such components and write them as a function of the Cartesian coordinates. This task is easy for a point charge:

$$E_x = K\frac{qx}{r^3} = K\frac{qx}{(x^2 + y^2 + z^2)^{3/2}}$$

$$E_y = K\frac{qy}{r^3} = K\frac{qy}{(x^2 + y^2 + z^2)^{3/2}} \tag{1.24}$$

$$E_z = K\frac{qz}{r^3} = K\frac{qz}{(x^2 + y^2 + z^2)^{3/2}}$$

To calculate the partial derivatives, it is quite useful to previously show that:

$$\frac{\partial}{\partial x}\left(\frac{x}{r^3}\right) = \frac{\dfrac{\partial x}{\partial x}r^3 - x\dfrac{\partial r^3}{\partial x}}{r^6} =$$

$$= \frac{r^3 - x\dfrac{\partial}{\partial x}\left(x^2 + y^2 + z^2\right)^{3/2}}{r^6} = r^{-3} - 3x^2 r^{-5} \qquad (1.25)$$

With the help of this derivative it is easy to calculate the divergence of the electric field:

$$\frac{\partial E_x}{\partial x} + \frac{\partial E_y}{\partial y} + \frac{\partial E_z}{\partial z} =$$

$$Kq\left(r^{-3} - 3x^2 r^{-5}\right) + Kq\left(r^{-3} - 3y^2 r^{-5}\right) + Kq\left(r^{-3} - 3z^2 r^{-5}\right) =$$

$$Kq\left(r^{-3} - 3x^2 r^{-5} + r^{-3} - 3y^2 r^{-5} + r^{-3} - 3z^2 r^{-5}\right) = \qquad (1.26)$$

$$Kq\left(3r^{-3} - 3r^{-5}(x^2 + y^2 + z^2)\right) =$$

$$Kq\left(3r^{-3} - 3r^{-5}r^2\right) = 0$$

This result reveals that the divergence of the electric field of a point charge is zero everywhere except at the origin of the coordinate system (where the charge is placed). Is this logical? Note that, according to Gauss's law, the divergence must be ρ/ε_0. As the charge is situated at the origin, the charge density is zero everywhere, with the exception of that point. Gauss's law is therefore valid at any point of space except the origin.

What happens at that singular point? On the one hand, the electric field tends to infinity, being not differentiable, and Equation 1.17 cannot be employed any more. If we try to evaluate the divergence by applying its definition, we conclude that the divergence also tends to infinity, because the flux remains constant as long as the surface encloses the point charge but the volume goes to zero. On the other hand, the charge density of a point charge tends to infinite as well. In any case, it should be kept in mind that point charges are only a simplified representation of reality.

In summary, in the case of a point charge, the divergence is nonzero only at the point where the charge is situated. But this is the point where the electric field lines begin (for positive charges) or terminate (for negative charges). Everywhere else, the divergence of the electric field is zero. We should therefore conclude that the electric field lines cannot begin or terminate on points where the divergence is zero. This idea is particularly applicable to the magnetic field, whose divergence is zero everywhere, $\vec{\nabla} \cdot \vec{B} = 0$, and its field lines cannot emanate from or terminate on any point. A vector field with zero divergence everywhere is called solenoidal.

The concept of divergence can also be applied to many other areas of physics and engineering, such as fluid dynamics. Let us consider the flow of a liquid, which can be described by means of a vector field of velocities. The points of positive divergence are sources (faucets) whereas the points or negative divergence are sinks (drains). At any other point, the divergence of the velocity field must be zero.

1.2.5. Example 2: Uniformly Charged Sphere

Let us now consider a more realistic case: a uniformly charged sphere, whose charge is q and radius is a. This sphere is centered at the origin of the coordinate system. Many textbooks of introduction to physics obtain the electric field of this charge distribution from the integral form of Gauss's law. The electric field created by the sphere is given by:

$$\vec{E} = K \frac{q}{r^2} \hat{r} = K \frac{q}{r^3} \vec{r} \qquad r \geq a$$

$$\vec{E} = K \frac{q\vec{r}}{a^3} \qquad r \leq a \tag{1.27}$$

Outside the sphere, the electric field is identical to that originated by a point charge at the origin, which verifies the differential form of Gauss's law (as proved in Example 1). But, can we say the same for the inner field?

First, we should identify the Cartesian components of the electric field. This is easy recalling that the vector position is $\vec{r} = x\hat{i} + y\hat{j} + z\hat{k}$. Thus:

$$E_x = K\frac{qx}{a^3}$$

$$E_y = K\frac{qy}{a^3} \qquad r \le a \tag{1.28}$$

$$E_z = K\frac{qz}{a^3}$$

Taking partial derivatives and adding gives:

$$\frac{\partial E_x}{\partial x} + \frac{\partial E_y}{\partial y} + \frac{\partial E_z}{\partial z} = \frac{Kq}{a^3} + \frac{Kq}{a^3} + \frac{Kq}{a^3} = 3\frac{Kq}{a^3} = 3\frac{q}{4\pi\varepsilon_0 a^3} \tag{1.29}$$

As can be seen, the divergence is not zero in this case because the region of space enclosed by the sphere is charged. Is this result identical to ρ/ε_0 ? As charge is uniformly distributed (inside the sphere), the charge density can be computed dividing the charge of the sphere by its volume:

$$\frac{\rho}{\varepsilon_0} = \frac{q/\frac{4}{3}\pi a^3}{\varepsilon_0} = \frac{3q}{4\pi\varepsilon_0 a^3} \tag{1.30}$$

As can be concluded, this value agrees with the divergence. Thus this electric field also satisfies Gauss's law in differential form.

1.3. AMPÈRE-MAXWELL'S LAW IN DIFFERENTIAL FORM

1.3.1. Circulation and Curl

First, we should recall the integral form of Ampère-Maxwell's law, because it will be our starting point:

$$\oint_C \vec{B} \cdot d\vec{l} = \mu_0\left(i_C + \varepsilon_0 \frac{d\Phi_E}{dt} \right) \tag{1.31}$$

where μ_0 is the magnetic permeability of empty space, i_C is the net current that passes through any surface bounded by the closed path C, and Φ_E is the flux of the electric field through the same surface. According to this law, the magnetic field has two sources: electric conduction currents and changing electric fields. The left side of this law is the so-called circulation of the magnetic field around the closed path. This line integral tells us if the tangential component of a vector field mostly points in the same direction as the infinitesimal displacements or in the opposite direction when we move along the path C. The concept of work can also be quite helpful to understand the meaning of this integral because the work done by a force along a closed path is no more than the circulation of such a force. We must bear in mind that the sign of the current is determined by the left-hand rule: if you wrap the fingers of your right hand around the path in the direction of integration, your thumb points in the direction of positive current. In any case, the complete analysis of the integral form of this law goes beyond the scope of this book. The readers interested in further details are referred to excellent textbooks of introductory level.

In the previous section, the divergence was found by considering the flux through an infinitesimal surface surrounding the point of interest. In this case, the curl at a given point P with coordinates (x, y, z) will be obtained by computing the circulation per unit area over an infinitesimal path around the specified point. The situation is sketched in Figure 1.2.

To begin with, we will consider the circulation over a rectangle centered at P and perpendicular to the x-axis. This rectangle has two sides parallel to the z-axis (sides 1 and 3), whose length is Δz, and two sides parallel to the y-axis (sides 2 and 4), whose length is Δy. Both dimensions are quite small because we are interested just in the neighborhood of P. Since this rectangle is a closed path, we must specify the direction of integration, i.e., the direction in which the rectangle is gone over, which is given by the wrapped fingers of your right hand when the thumb points to the x-direction.

At any rate, the line integral along the rectangle can be split into four line integrals along each of the sides:

$$\oint_C \vec{B} \cdot d\vec{l} = \int_{C1} \vec{B} \cdot d\vec{l} + \int_{C2} \vec{B} \cdot d\vec{l} + \int_{C3} \vec{B} \cdot d\vec{l} + \int_{C4} \vec{B} \cdot d\vec{l} \qquad (1.32)$$

First, we will estimate the line integral along side 1 (C1) by assuming that the magnetic field there can be approximated by the magnetic field at the center of the side $\vec{B}_1 \equiv \vec{B}(x, y + \Delta y/2, z)$. Taking into account that the displacement along this side is $\Delta \vec{l}_1 = \Delta z \hat{k}$, we can approximately write:

$$\int_{C1} \vec{B} \cdot d\vec{l} \cong \vec{B}_1 \cdot \Delta \vec{l}_1 = \vec{B}_1 \cdot \hat{k} \Delta z = \vec{B}(x, y + \Delta y/2, z) \cdot \hat{k} \Delta z =$$
$$= B_z(x, y + \Delta y/2, z) \cdot \Delta z \tag{1.33}$$

Now let us consider side 3 (the opposite one), for which $\vec{B}_3 \equiv \vec{B}(x, y - \Delta y/2, z)$ and $\Delta \vec{l}_3 = -\Delta z \hat{k}$. Then, the line integral is approximately given by :

$$\int_{C3} \vec{B} \cdot d\vec{l} \cong \vec{B}_3 \cdot \Delta \vec{l}_3 = \vec{B}_3 \cdot (-\hat{k} \Delta z) = -\vec{B}(x, y - \Delta y/2, z) \cdot \hat{k} \Delta z =$$
$$= -B_z(x, y - \Delta y/2, z) \cdot \Delta z \tag{1.34}$$

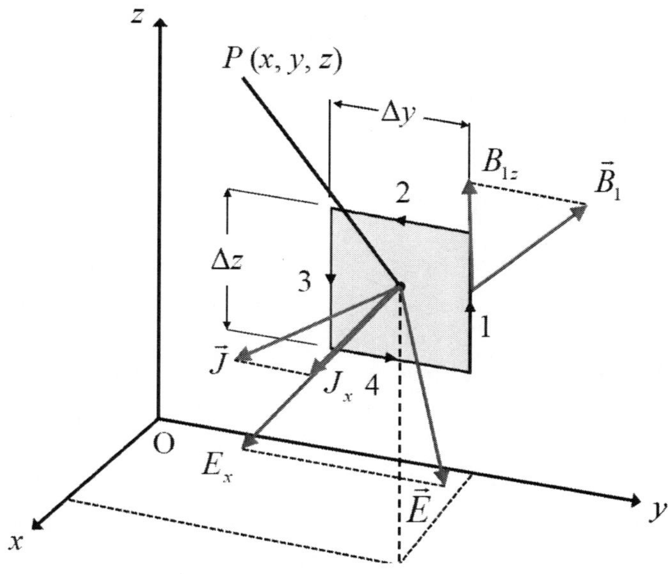

Figure 1.2. Rectangle of small dimensions centered at P. The magnetic field at one of its sides and the component parallel to this side are also shown.

Thus the line integral along these two opposite sides is:

$$\int_{C1}\vec{B}\cdot d\vec{l} + \int_{C3}\vec{B}\cdot d\vec{l} \cong \left[B_z(x,y+\tfrac{1}{2}\Delta y,z) - B_z(x,y-\tfrac{1}{2}\Delta y,z)\right]\Delta z \quad (1.35)$$

Doing likewise for sides 2 and 4, we obtain:

$$\int_{C2}\vec{B}\cdot d\vec{l} + \int_{C4}\vec{B}\cdot d\vec{l} \cong \left[-B_y(x,y,z+\tfrac{1}{2}\Delta z) + B_y(x,y,z-\tfrac{1}{2}\Delta z)\right]\Delta y \quad (1.36)$$

Thus the circulation around the rectangle is finally given by:

$$\oint_{C_{zy}}\vec{B}\cdot d\vec{l} \cong \left[B_z(x,y+\tfrac{1}{2}\Delta y,z) - B_z(x,y-\tfrac{1}{2}\Delta y,z)\right]\Delta z$$
$$-\left[B_y(x,y,z+\tfrac{1}{2}\Delta z) - B_y(x,y,z-\tfrac{1}{2}\Delta z)\right]\Delta y \quad (1.37)$$

where C_{zy} denotes a closed path parallel to the zy-plane or perpendicular to the x-axis (the rectangle in this case). As can be seen, this circulation vanishes when the rectangle dimensions tend to zero. For this reason, let us define the x-component of the curl as the circulation *per unit area* around an infinitesimal closed path surrounding the point of interest and bounding a surface perpendicular to the x-axis:

$$\left(rot\vec{B}\right)_x \equiv \lim_{A(C_{zy})\to 0} \frac{\oint_{C_{zy}}\vec{B}\cdot d\vec{l}}{A} \quad (1.38)$$

where $A(C_{zy})$ stands for the area of C_{zy}. The other Cartesian components can be defined in a similar way. Dividing the circulation by the area of this rectangle, $A = \Delta y\Delta z$, and taking the limits $\Delta y \to 0, \Delta z \to 0$ yields:

$$\left(rot\vec{B}\right)_x = \lim_{A(C_{zy})\to 0} \frac{\oint_{C_{zy}}\vec{B}\cdot d\vec{l}}{A} =$$
$$= \lim_{\Delta y\to 0} \frac{\left[B_z(x,y+\tfrac{1}{2}\Delta y,z) - B_z(x,y-\tfrac{1}{2}\Delta y,z)\right]}{\Delta y}$$
$$- \lim_{\Delta z\to 0} \frac{\left[B_y(x,y,z+\tfrac{1}{2}\Delta z) - B_y(x,y,z-\tfrac{1}{2}\Delta z)\right]}{\Delta z} \quad (1.39)$$

Again, the limits appearing in this equation can be identified as partial derivatives, which gives:

$$\left(rot\vec{B}\right)_x = \lim_{A(C_{zy})\to 0} \frac{\oint_{C_{zy}} \vec{B}\cdot d\vec{l}}{A} = \left(\frac{\partial B_z}{\partial y} - \frac{\partial B_y}{\partial z}\right) \tag{1.40}$$

Doing the same for the y- and z-component, we finally obtain:

$$\left(rot\vec{B}\right)_x = \lim_{A(C_{zy})\to 0} \frac{\oint_{C_{zy}} \vec{B}\cdot d\vec{l}}{A} = \left(\frac{\partial B_z}{\partial y} - \frac{\partial B_y}{\partial z}\right)$$

$$\left(rot\vec{B}\right)_y = \lim_{A(C_{zx})\to 0} \frac{\oint_{C_{xz}} \vec{B}\cdot d\vec{l}}{A} = \left(\frac{\partial B_x}{\partial z} - \frac{\partial B_z}{\partial x}\right) \tag{1.41}$$

$$\left(rot\vec{B}\right)_z = \lim_{A(C_{xy})\to 0} \frac{\oint_{C_{xy}} \vec{B}\cdot d\vec{l}}{A} = \left(\frac{\partial B_y}{\partial x} - \frac{\partial B_x}{\partial y}\right)$$

The curl can be calculated from these scalar expressions if the functional forms of the Cartesian components of the magnetic field, $B_x(x,y,z)$, $B_y(x,y,z)$ and $B_z(x,y,z)$ are known. This result can also be summarized in a vector equation:

$$rot\vec{B} = \left(\frac{\partial B_z}{\partial y} - \frac{\partial B_y}{\partial z}\right)\hat{i} + \left(\frac{\partial B_x}{\partial z} - \frac{\partial B_z}{\partial x}\right)\hat{j} + \left(\frac{\partial B_y}{\partial x} - \frac{\partial B_x}{\partial y}\right)\hat{k} \tag{1.42}$$

Before reconsidering Ampère-Maxwell's equation from this perspective, let us try to visualize how the curl works with the help of Figure 1.3, which shows a sketch of a two-dimensional velocity field around a point of the xy-plane $P(x, y, 0)$. The velocities acting on four points of the neighborhood of P (A, B, C and D) are also drawn. Looking first at the vertical velocities, it can be concluded that $\partial v_y / \partial x$ must be positive because the y-component of velocity increases from C to D. Imagine a tiny paddle wheel held at the point of interest. These two velocities would make it rotate counterclockwise, the positive direction of rotation according to the right-hand rule if the thumb points in the +z-direction. Looking now at the x-component, it can be seen that v_x is greater above P, so $\partial v_x / \partial y$ is also positive. However, the couple of horizontal velocities would make the paddle wheel rotate clockwise, the negative direction of rotation. For this reason, this partial derivative is subtracted from the previous one. This example also suggests that each component of the curl tells us the tendency of the field to rotate in one of the coordinate planes. For instance, if the curl of a field at given point has a large z-component, the field will mostly circulate around that point in the xy-plane (with a direction of circulation given by the right-hand rule).

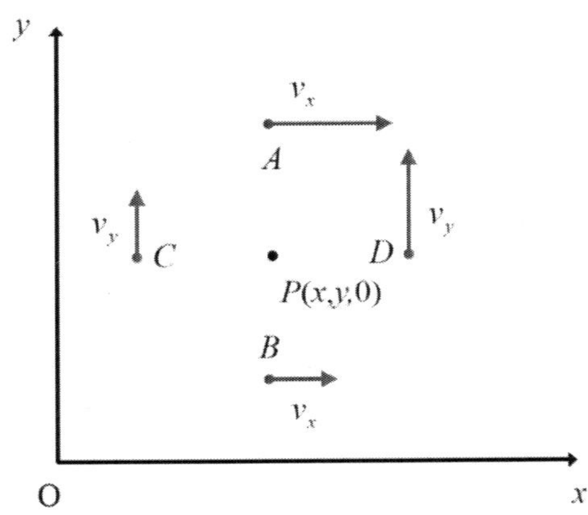

Figure 1.3. Sketch of a two-dimensional velocity field around a point of the xy-plane P. The vertical velocities would make a tiny paddle wheel rotate counterclockwise whereas the horizontal velocities would make it rotate clockwise.

We can take this mechanical analogy further: In a fluid that is rotating like a rigid body, the curl of the velocity vector is twice the angular velocity vector (see Problem 1.7). In other words: the angular velocity of a paddle wheel would be half of the curl of the velocity field.

1.3.2. The Curl and Ampère-Maxwell's Law

In the previous subsection, we derived expressions for calculating the curl of the magnetic field (or any other vector field) from its Cartesian components. On the other hand, the curl can also be calculated from Ampère-Maxwell's equation. First, let us consider the x-component:

$$(rot\vec{B})_x = \lim_{A(C_{zy})\to 0} \frac{\oint_{C_{zy}}\vec{B}\cdot d\vec{l}}{A} = \lim_{A(C_{zy})\to 0} \frac{\mu_0\left(i_C + \varepsilon_0 \dfrac{d\Phi_E}{dt}\right)}{A} \qquad (1.43)$$

At this point, we recall that the current can be considered as the flux of the current density ($\vec{J}$). As the surface bounded by the path C_{zy} is perpendicular to the x-axis, the only component of the current density contributing to the electric current is J_x. In other words: $i_C = J_x A$. Likewise, the only component of the electric field contributing to the flux through the surface bounded by C_{zy} is E_x, which implies that $\Phi_E = E_x A$. Inserting these two equations into 1.43, we have:

$$(rot\vec{B})_x = \lim_{A(C_{zy})\to 0} \frac{\oint_{C_{zy}}\vec{B}\cdot d\vec{l}}{A} = \lim_{S(C_{zy})\to 0} \frac{\mu_0\left(J_x A + \varepsilon_0 \dfrac{d(E_x A)}{dt}\right)}{A} \qquad (1.44)$$

Since A does not depend on time, the area can be come out of the time derivative and after factoring out:

$$(rot\vec{B})_x = \lim_{A(C_{zy})\to 0} \frac{\oint_{C_{zy}}\vec{B}\cdot d\vec{l}}{A} =$$

$$= \lim_{A(C_{zy})\to 0} \frac{\mu_0\left(J_x + \varepsilon_0 \dfrac{\partial E_x}{\partial t}\right)A}{A} = \qquad (1.45)$$

$$= \mu_0\left(J_x + \varepsilon_0 \frac{\partial E_x}{\partial t}\right)$$

Note that the time derivative of the x-component of electric field is partial because this quantity not only depends on time but also on x, y and z. However, the time derivative included in the integral form of Ampère-Maxwell law (see Equations 1.31 and 1.43) is not partial because, in this form, the flux is usually evaluated over relatively large surfaces and not at a given point. Under such circumstances, the flux only depends on time.

Doing the same for the y- and z-component, we complete the set of three Cartesian components of the curl:

$$(rot\vec{B})_x = \mu_0\left(J_x + \varepsilon_0\frac{\partial E_x}{\partial t}\right)$$

$$(rot\vec{B})_y = \mu_0\left(J_y + \varepsilon_0\frac{\partial E_y}{\partial t}\right) \tag{1.46}$$

$$(rot\vec{B})_z = \mu_0\left(J_z + \varepsilon_0\frac{\partial E_z}{\partial t}\right)$$

Comparing Equations 1.46 and 1.41 gives:

$$\left(\frac{\partial B_z}{\partial y} - \frac{\partial B_y}{\partial z}\right) = \mu_0\left(J_x + \varepsilon_0\frac{\partial E_x}{\partial t}\right)$$

$$\left(\frac{\partial B_x}{\partial z} - \frac{\partial B_z}{\partial x}\right) = \mu_0\left(J_y + \varepsilon_0\frac{\partial E_y}{\partial t}\right) \tag{1.47}$$

$$\left(\frac{\partial B_y}{\partial x} - \frac{\partial B_x}{\partial y}\right) = \mu_0\left(J_z + \varepsilon_0\frac{\partial E_z}{\partial t}\right)$$

These are the three scalar equations we were looking for. They relate the Cartesian components of the electric field at a given point to the current density and electric field at the point of interest.

1.3.3. A Compact Differential Form for Ampère-Maxwell's Law

These equations can be written in a more compact form, which is also much easier to memorize, with the help of nabla, the vector operator. First, let us calculate the cross (or vector) product of nabla and the magnetic field:

$$\vec{\nabla}\times\vec{B} \equiv \begin{vmatrix} \hat{i} & \hat{j} & \hat{k} \\ \frac{\partial}{\partial x} & \frac{\partial}{\partial y} & \frac{\partial}{\partial z} \\ B_x & B_y & B_z \end{vmatrix} = \tag{1.48}$$

$$= \left(\frac{\partial B_z}{\partial y} - \frac{\partial B_y}{\partial z}\right)\hat{i} + \left(\frac{\partial B_x}{\partial z} - \frac{\partial B_z}{\partial x}\right)\hat{j} + \left(\frac{\partial B_y}{\partial x} - \frac{\partial B_x}{\partial y}\right)\hat{k}$$

This result is identical to Equation 1.42. Thus the curl can be seen as the cross product of nabla and the magnetic field. Consequently:

$$\vec{\nabla} \times \vec{B} = \mu_0 \left(J_x + \varepsilon_0 \frac{\partial E_x}{\partial t} \right) \hat{i} + \mu_0 \left(J_y + \varepsilon_0 \frac{\partial E_y}{\partial t} \right) \hat{j} + \left(J_z + \varepsilon_0 \frac{\partial E_z}{\partial t} \right) \hat{k}$$

$$(1.49)$$

Or, more compactly:

$$\vec{\nabla} \times \vec{B} = \mu_0 \left(\vec{J} + \varepsilon_0 \frac{\partial \vec{E}}{\partial t} \right) \tag{1.50}$$

If the differential and integral forms of Ampère-Maxwell's law are compared, it can be concluded that:

1) The circulation of the magnetic field (a scalar quantity) has been replaced by the curl (vector).
2) The current (scalar) has been replaced by the current density (vector).
3) The time derivative of the flux of the electric field has been replaced by the time derivative of this vector.

1.3.4. Example: Magnetic Field of a Current-Carrying Long Straight Wire

Imagine a long straight wire (whose radius is negligible) carrying a steady current I along the z-axis (in the $+z$-direction, see Figure 1.4) and consider a point $P(x,y,z)$ outside the wire.

The magnetic field due to this wire at P is usually calculated in many textbooks from Biot-Sarvat's law and/or the integral form of Ampère-Maxwell's law: $B = \mu_0 I / (2\pi r)$, where r is the distance from the wire to P.

Here, we will prove that the magnetic field originated by this wire satisfies the differential form of Ampère-Maxwell's law. First, note that r does not depend on z and $r = \sqrt{x^2 + y^2}$. From Figure 1.4, it is easy to show that:

$$B_x = -B\sin\varphi = -\frac{\mu_0 I}{2\pi r}\frac{y}{r} = -\frac{\mu_0 I y}{2\pi r^2}$$

$$B_y = B\cos\varphi = \frac{\mu_0 I}{2\pi r}\frac{x}{r} = \frac{\mu_0 I x}{2\pi r^2} \qquad (1.51)$$

$$B_z = 0$$

The curl $\vec{\nabla}\times\vec{B}$ can be calculated from these components by means of the determinant:

$$\vec{\nabla}\times\vec{B} \equiv \begin{vmatrix} \hat{i} & \hat{j} & \hat{k} \\ \frac{\partial}{\partial x} & \frac{\partial}{\partial y} & \frac{\partial}{\partial z} \\ B_x & B_y & B_z \end{vmatrix} \qquad (1.52)$$

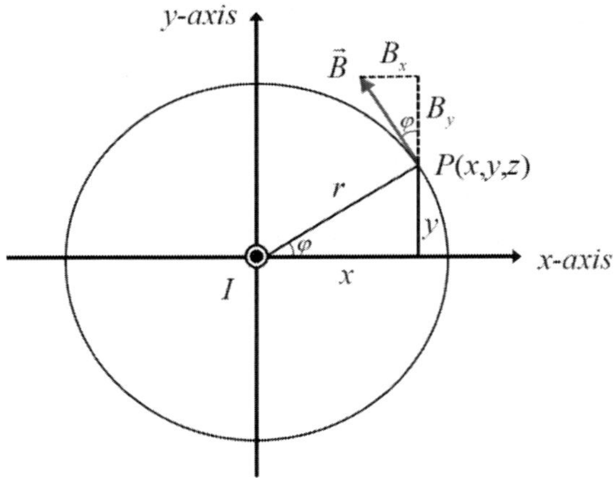

Figure 1.4. Magnetic field created by a long straight wire carrying a steady current I along the z-axis (in the $+z$-direction) at a point P. The Cartesian components of this field are also drawn.

Before expanding this determinant, it should be mentioned that:1) The z-component of the electric field is zero ($B_z = 0$); 2) None of the components depends on z, so the partial derivative with respect to this coordinate will be zero. Symbolically, this fact can be accounted for replacing $\partial/\partial z$ by 0. Thus:

$$\vec{\nabla} \times \vec{B} \equiv \begin{vmatrix} \hat{i} & \hat{j} & \hat{k} \\ \frac{\partial}{\partial x} & \frac{\partial}{\partial y} & 0 \\ B_x & B_y & 0 \end{vmatrix} = \left(\frac{\partial B_y}{\partial x} - \frac{\partial B_x}{\partial y} \right) \hat{k} \tag{1.53}$$

In addition:

$$\frac{\partial B_y}{\partial x} = \frac{\partial}{\partial x} \left(\frac{\mu_0 I x}{2\pi r^2} \right) = \frac{\mu_0 I}{2\pi} \frac{\partial}{\partial x} \left(\frac{x}{r^2} \right) =$$

$$= \frac{\mu_0 I}{2\pi} \frac{\frac{\partial x}{\partial x} r^2 - x \frac{\partial r^2}{\partial x}}{r^4} = \frac{\mu_0 I}{2\pi} \frac{r^2 - 2x^2}{r^4} \tag{1.54}$$

Analogously:

$$\frac{\partial B_x}{\partial y} = \frac{\partial}{\partial y} \left(-\frac{\mu_0 I y}{2\pi r^2} \right) = -\frac{\mu_0 I}{2\pi} \frac{\partial}{\partial y} \left(\frac{y}{r^2} \right) = -\frac{\mu_0 I}{2\pi} \frac{r^2 - 2y^2}{r^4} \tag{1.55}$$

And finally:

$$\vec{\nabla} \times \vec{B} = \left(\frac{\partial B_y}{\partial x} - \frac{\partial B_x}{\partial y} \right) \hat{k} =$$

$$\left(\frac{\mu_0 I}{2\pi} \frac{r^2 - 2x^2}{r^4} + \frac{\mu_0 I}{2\pi} \frac{r^2 - 2y^2}{r^4} \right) \hat{k} = \tag{1.56}$$

$$\frac{\mu_0 I}{2\pi} \frac{2r^2 - 2(x^2 + y^2)}{r^4} \hat{k} = 0$$

We therefore conclude that the curl is zero at any point outside the wire. For such a point $\vec{J} = 0$ because there is not electric current outside the wire. In addition, the electric field is zero everywhere. Consequently:

$$\mu_0\left(\vec{J} + \varepsilon_0\,\frac{\partial \vec{E}}{\partial t}\right) = 0 \tag{1.57}$$

Thus the magnetic field of a long straight wire carrying a steady current satisfies Ampère-Maxwell law outside the wire. What happens along the wire? In this case, the curl cannot be computed from partial derivatives because the field tends to infinity when r approaches 0.

1.4. FARADAY'S LAW IN DIFFERENTIAL FORM

Faraday's law tells us that changing the magnetic flux through a surface bounded by a wire induces an electromotive force in the wire.

Its integral form is:

$$\oint_C \vec{E}\cdot d\vec{l} = -\frac{d\Phi_B}{dt} \tag{1.58}$$

As can be seen, the left side of this equation is the circulation of the electric field around the wire, which is no more than the electromotive force. There is certain analogy between the integral forms of Faraday's law and Ampère-Maxwell law. As can be seen, both involved the circulation of a vector field on the left side and the time derivative of a flux on the right side. As mentioned in section 1.3.3, the circulation was replaced by the curl and the flux was replaced by the corresponding field when the differential form of Ampère-Maxwell's law was derived. The same approach applied to the integral form of Faraday's law for magnetic induction leads to:

$$\vec{\nabla}\times\vec{E} = -\frac{\partial \vec{B}}{\partial t} \tag{1.59}$$

The physical meaning of this equation is essentially the same: Variable magnetic fields are sources of the electric field. It should be stressed, however, that such induced electric fields have field lines that loop back on themselves, with no points of origination or termination. On the other hand, the electric field originated by static distributions of charges is conservative, which means that $\vec{\nabla} \times \vec{E} = 0$.

1.5. SOME CONCLUDING REMARKS

To end this chapter, Maxwell's equations in differential form are summarized:

$$\vec{\nabla} \cdot \vec{E} = \rho / \varepsilon_0$$

$$\vec{\nabla} \times \vec{E} = -\frac{\partial \vec{B}}{\partial t}$$

$$\vec{\nabla} \cdot \vec{B} = 0 \qquad\qquad (1.60)$$

$$\vec{\nabla} \times \vec{B} = \mu_0 \left(\vec{J} + \varepsilon_0 \frac{\partial \vec{E}}{\partial t} \right)$$

The left side of these equations always tells us the field we are talking about whereas the right side informs about its sources. As can be seen, each field has two equations associated to it. One of them relates the divergence of the field of interest to sources of scalar nature. The other relates the curl of the field to sources of vector nature. Charge is the scalar source of electric fields. As the 'magnetic charge' does not exist, the divergence of the magnetic field is zero. Changing magnetic fields are the vector source of the electric field, whereas electric currents and time-varying electric fields are the vector sources of the magnetic field.

In empty space, there cannot be charges or electric currents. Thus the divergence of the electric field is zero and its field lines do not have points of origination or termination. However, there might be variable fields. Be careful in applying Faraday's law and Ampère-Maxwell's law.

Although this book is restricted to problems in empty space, Maxwell's equations in the form written in Equations 1.60 can be theoretically applied to electric and magnetic fields in matter provided that bound charges and currents are accounted for. However, this is not a trivial matter, because they are not known a priori and depend on electric and magnetic fields. For this reason, Gauss's law for the electric field and Ampère-Maxwell's law have versions (outlined in appendixes of this chapter) that only depend on free charges and currents. Gauss's law for magnetism and Faraday's law do not require 'matter-adapted' versions because charges or currents are not included in their sources.

Before ending this chapter, here is an important rule of thumb concerning the application of Maxwell's equations: Maxwell's equations in differential form are applied to individual points and not to closed surfaces or paths. Just as physical quantities defined from time derivatives refer to a given instant (e.g., instantaneous velocity), properties based on space derivatives refer to points.

APPENDIX 1.1: GAUSS'S LAW IN DIELECTRICS

Gauss's law as presented in a previous section can be applied to the electric field whether in free space or in matter. However, we should keep in mind that the charge enclosed by a Gaussian surface (in the integral form) and the charge density of the right side of the differential form include both free and bound charge. Recall that bound charges are slightly displaced by the electric field but cannot move freely through the material. When a dielectric is placed inside an external electric field, its molecules are polarized in the direction of this field, which produces a net dipole moment parallel to the applied electric field. As a result of this polarization, certain net bound charge can appear in some regions of the dielectric. We will refer to this bound charge as polarization charge. Dealing with polarization charge is not a trivial task because it depends on the resulting electric field but the electric field depends in turn on the total charge (included polarization charge). In this section we will outline a new version of Gauss's law that does not depend on polarization charge. A rigorous treatment of this issue goes beyond the scope of this book and can be found in advanced textbooks.

Let us imagine a rectangular prism with edges parallel to the coordinate axes inside a dielectric and an electric field forming an angle ϕ with the x-axis (see Figure 1.5).

Now consider a molecule inside the prism. Its positive and negative charge centers (with charges $+q$ and $-q$) are separated by an average distance L as a result of the electric field, so its dipole moment is $\vec{p} = q\vec{L}$. This dipole moment is parallel to the electric field. If this molecule is far enough from the surface of this prism, it does no contribute to a net bound charge inside it because both charge centers (positive and negative) are included in the prism. But let us imagine another molecule near the right face of the prism. If the distance between the center of negative charge and the surface is less than $L\cos\phi$, then the center of the positive charge is out of the prism surface and this dipole therefore contributes to the net charge inside. All the dipoles having their centers of negative charge inside a layer adjacent to the right face whose thickness is $L\cos\phi$ also contribute to bound charge. If the number of molecules per unit volume (N) is known the number of these dipoles can be calculated multiplying N by the volume of this layer, $AL\cos\phi$, where A is the area of the right face of the prism. Each one of these dipoles contributes with $-q$ to the net bound charge inside the prism.

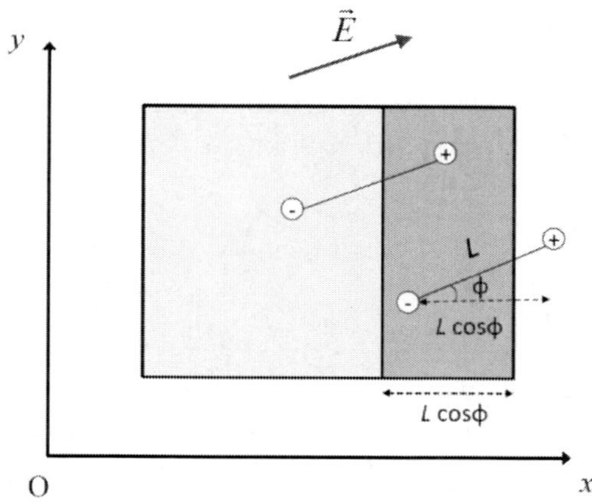

Figure 1.5. Projection of a rectangular prism with edges parallel to the coordinate axes inside a dielectric. Two electric dipoles are also drawn. The inner one does not contribute to the net bound charge inside the prism. However, the dipoles whose center of negative charge is inside the shaded region do contribute to the net bound charge. The dimensions of the prism are not drawn to scale.

Thus, the right face of this figure contributes with a bound charge:

$$Q_b = -qNAL\cos\phi = -NqLA\cos\phi \tag{1.61}$$

Recalling that $p = qL$ this expression can be written as:

$$Q_b = -NpA\cos\phi = -N\vec{p}\cdot\vec{A} \tag{1.62}$$

Defining the polarization vector as the net dipole moment per unit volume, $\vec{P} = N\vec{p}$, we end up with:

$$Q_b = -\vec{P} \cdot \vec{A} \tag{1.63}$$

The bound charge inside the prism is obtained adding in the contributions of the rest of faces:

$$Q_b = -\sum_i \vec{P} \cdot \vec{A}_i \tag{1.64}$$

This result can be easily generalized for an arbitrary closed surface S and a polarization vector that varies from point to point considering infinitesimal area vectors and integrating over S:

$$Q_b = -\oint_S \vec{P} \cdot d\vec{A} \tag{1.65}$$

This expression tells us that the bound charge enclosed by a surface S is given by the flux of the polarization vector through such a surface, but with opposite sign. Then, the volume density of bound charge, ρ_b, can be computed as:

$$\rho_b = \lim_{V \to 0} \frac{Q_b}{V} = -\lim_{V \to 0} \frac{\oint_S \vec{P} \cdot d\vec{A}}{V} = -\vec{\nabla} \cdot \vec{P} \tag{1.66}$$

Both results can help us to transform Gauss's law into a useful version for dielectrics. For instance, let us consider the differential form:

$$\vec{\nabla} \cdot \vec{E} = \frac{\rho_f + \rho_b}{\varepsilon_0} \tag{1.67}$$

Here, we have explicitly distinguished between free charge, ρ_f, which is usually known, and the bound charge, ρ_b, which is not generally known *a priori*. Inserting Equation 1.66 into Gauss's law gives:

$$\vec{\nabla} \cdot \vec{E} = \frac{\rho_f - \vec{\nabla} \cdot \vec{P}}{\varepsilon_0} \qquad (1.68)$$

which can also be written as:

$$\varepsilon_0 \vec{\nabla} \cdot \vec{E} + \vec{\nabla} \cdot \vec{P} = \rho_f \qquad (1.69)$$

Taking into account that the divergence is a linear operator leads to:

$$\vec{\nabla} \cdot \left(\varepsilon_0 \vec{E} + \vec{P} \right) = \rho_f \qquad (1.70)$$

The term in parentheses is often written as a vector called electric displacement:

$$\vec{D} \equiv \varepsilon_0 \vec{E} + \vec{P} \qquad (1.71)$$

In terms of the electric displacement, Gauss's law is expressed as:

$$\vec{\nabla} \cdot \vec{D} = \rho_f \qquad (1.72)$$

Note that the divergence of the electric displacement does not depend on bound charge. But the price of this achievement is introducing a new vector in the treatment of electromagnetic problems. Luckily, there are many dielectrics for which the electric field and the electric displacement are proportional:

$$\vec{D} = \varepsilon \vec{E} \qquad (1.73)$$

where ε is a constant quantity known as the permittivity of the dielectric, which considerably facilitates the handling of situations where dielectrics are involved. If this property is constant the constitutive relation 1.73 can be inserted into Gauss's law, which gives:

$$\vec{\nabla} \cdot \vec{E} = \frac{\rho_f}{\varepsilon} \qquad (1.74)$$

which means that Gauss's law can be applied inside dielectrics that obey this constitutive relation replacing the permittivity of empty space by the permittivity of the dielectrics (providing that this property is constant). In this book, however, we will restrict ourselves to problems in empty space.

APPENDIX 1.2: AMPÈRE-MAXWELL'S LAW IN MAGNETIC MATERIALS

Ampère-Maxwell's law also requires a modified version that can be easily applied in matter because the current enclosed by a path in its integral form and the density current appearing in the differential form include both free and bound currents. When a piece of material is placed inside an external magnetic field, its molecules or atoms acquire a net magnetic dipole moment, which can be visualized as a microscopic current loop, whose area is A_{loop}. As a result of this magnetization, net bound currents can appear in some regions of this material. We will refer to these bound currents as magnetization currents. As the reader can guess, magnetization currents are not known *a priori* and dealing with them becomes a trouble because they are sources of the magnetic field but this field depends in turn on them. In this section we will derive a new version of Ampère-Maxwell's law that does not depend on magnetization currents.

Let us imagine a rectangle with sides parallel to the xy-plane inside a magnetic material whose atoms have acquired on average a net magnetic dipole moment forming an angle ϕ with the y-axis (see Figure 1.6). The associated magnetic current loops that cross this rectangle far enough from its edges do not contribute to a net magnetization current through it because each loop crosses the rectangle twice with currents in opposite directions. But let us imagine a microscopic magnetic dipole near the right side of the rectangle. If its surface is crossed by the side of the square, then this loop does contribute to a net bound current through the rectangle. Note that all the loops that have their left sides in a rectangular prism near the right side (shaded in Figure 1.6), whose cross-sectional area is $A_{loop}\cos\phi$, also contribute to this net bound current. If the number of molecules per unit volume (N) is known, the number of these loops can be calculated multiplying N by the volume of this prism, $A_{loop}l\cos\phi$, where l is the length of the right side of the rectangle. Each one of these magnetic dipoles contributes with I_{loop} to the net bound current inside the rectangle. Thus, the right side of this figure contributes with a bound current:

$$I_b = I_{loop}NA_{loop}l\cos\phi = NI_{loop}A_{loop}l\cos\phi \qquad (1.75)$$

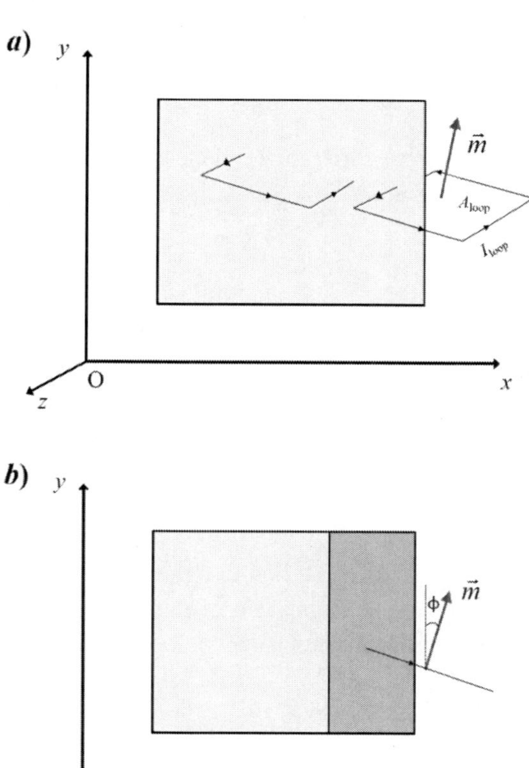

Figure 1.6. a) Rectangle with sides parallel to the *xy*-plane inside a magnetic material. Two microscopic amperian currents are also drawn. The inner one does not contribute to the net bound current through the rectangle. However, the amperian current near the right side does contribute to the net bound current. b) Projection on the *xy*-plane. In both cases, the rectangle dimensions are not drawn to scale.

Recalling that the magnetic dipole moment is given by $m = I_{loop}A_{loop}$ this expression can be written as:

$$I_b = Nml\cos\phi = N\vec{m}\cdot\vec{l} \qquad (1.76)$$

Defining the magnetization vector as the net magnetic dipole moment per unit volume, $\vec{M} = N\vec{m}$, leads to:

$$I_b = \vec{M} \cdot \vec{l} \tag{1.77}$$

The total bound current through the rectangle is obtained adding in the contributions of the rest of sides:

$$I_b = \sum_i \vec{M} \cdot \vec{l}_i \tag{1.78}$$

This result can be easily generalized for an arbitrary closed curve C and a magnetization vector that varies from point to point considering infinitesimal displacement vectors and integrating around C:

$$I_b = \oint_C \vec{M} \cdot d\vec{l} \tag{1.79}$$

This expression tells us that the bound current that penetrates any surface bounded by C is given by the circulation of the magnetization vector around the path C. Thus the bound current density can be computed as:

$$J_{bx} = \lim_{A(C_{zy}) \to 0} \frac{\oint_{C_{zy}} \vec{M} \cdot d\vec{l}}{A} = \left(\vec{\nabla} \times \vec{M} \right)_x$$

$$J_{by} = \lim_{A(C_{xz}) \to 0} \frac{\oint_{C_{xz}} \vec{M} \cdot d\vec{l}}{A} = \left(\vec{\nabla} \times \vec{M} \right)_y \tag{1.80}$$

$$J_{bz} = \lim_{A(C_{xy}) \to 0} \frac{\oint_{C_{xy}} \vec{M} \cdot d\vec{l}}{A} = \left(\vec{\nabla} \times \vec{M} \right)_z$$

These three scalar equations can be expressed in a more compact vector form as:

$$\vec{J}_b = \vec{\nabla} \times \vec{M} \tag{1.81}$$

Equations 1.79 and 1.81 can help us to transform Ampère-Maxwell's law into a useful version for magnetic materials. For instance, let us consider the differential form:

$$\vec{\nabla} \times \vec{B} = \mu_0 \left(\vec{J}_f + \vec{J}_b + \vec{J}_P + \varepsilon_0 \frac{\partial \vec{E}}{\partial t} \right) \tag{1.82}$$

Here, we have explicitly distinguished between free currents ($\vec{J}_f$), which are usually known, and bound currents ($\vec{J}_b$), which are not generally known *a priori*. We have also accounted for polarization currents, ($\vec{J}_P$) originated by time-varying electric dipoles since any movement of charge constitutes an electric current. It can be shown that the polarization current density is:

$$\vec{J}_P = \frac{\partial \vec{P}}{\partial t} \tag{1.83}$$

Inserting Equations 1.83 and 1.81 into Ampère-Maxwell's law gives:

$$\vec{\nabla} \times \vec{B} = \mu_0 \left(\vec{J}_f + \vec{\nabla} \times \vec{M} + \frac{\partial \vec{P}}{\partial t} + \varepsilon_0 \frac{\partial \vec{E}}{\partial t} \right) \tag{1.84}$$

After some algebra, the above expression becomes:

$$\frac{1}{\mu_0} \vec{\nabla} \times \vec{B} - \vec{\nabla} \times \vec{M} = \left(\vec{J}_f + \frac{\partial \vec{P}}{\partial t} + \varepsilon_0 \frac{\partial \vec{E}}{\partial t} \right) \tag{1.85}$$

Taking into account that the curl and partial derivatives are linear operators leads to:

$$\vec{\nabla} \times \left(\frac{1}{\mu_0} \vec{B} - \vec{M} \right) = \vec{J}_f + \frac{\partial \left(\vec{P} + \varepsilon_0 \vec{E} \right)}{\partial t} \tag{1.86}$$

The right side can be expressed in a more compact form recalling the definition of electric displacement (see previous section):

$$\vec{\nabla} \times \left(\frac{1}{\mu_0} \vec{B} - \vec{M} \right) = \vec{J}_f + \frac{\partial \vec{D}}{\partial t} \tag{1.87}$$

The term in parentheses is often denoted as $\vec{H}$ but physicists do not reach an agreement about its name. For this reason, here it will be called the H-field:

$$\vec{H} \equiv \frac{1}{\mu_0} \vec{B} - \vec{M} \tag{1.88}$$

In terms of the H-field, Ampère-Maxwell's law is expressed as:

$$\vec{\nabla} \times \vec{H} = \vec{J}_f + \frac{\partial \vec{D}}{\partial t} \tag{1.89}$$

Note that the curl of the H-field does not depend on bound currents. But we have had to introduce a new vector in the treatment of electromagnetic problems. Luckily, there are many materials for which B-field and H-field are proportional:

$$\vec{H} = \vec{B} / \mu \tag{1.90}$$

where μ is a quantity known as magnetic permeability. If this property is constant the constitutive relation (Equation 1.90) can be inserted into Ampère-Maxwell's law, which gives:

$$\vec{\nabla} \times \vec{B} = \mu \left(\vec{J}_f + \frac{\partial \vec{D}}{\partial t} \right) \tag{1.91}$$

which means that this law can be applied in materials that obey this constitutive relation replacing the permeability of empty space by its permeability (providing that this property is constant). What is more, there are many materials for which the relative difference between μ and μ_0 is extremely low (or the order of 10^{-5} for many solids and 10^{-9} for gases).

HISTORICAL NOTE: THE CREATIVE PROCESS IN PHYSICS. MAXWELL AND HIS EQUATIONS

The movies and fiction make common people get the idea that the major contributions to physics are made by unusual, outstanding and brilliant individuals. People also think that most of their inventions and discoveries have arisen in isolated moments of special inspiration. According to the stereotype of physicist, these scientists are absorbed, disheveled and unsociable, whether they are genius or not. They are not concerned about the matters of the average citizen. In short, physicists live in a world of their own, are a bit crazy and can fill up several sheets of paper with completely incomprehensible formulas if you dare to ask them what things come into their mind. A very famous example of this kind of stereotype is likely to be Isaac Newton. Many people believe that Newton suddenly thought of the theory of universal gravitation just after seeing an apple fall, but this is no more than a too simplistic (and wrong) version of some information spread by his biographer and friend William Stukeley.

Reality is very different. The creative process in science, and more specifically in physics, is slow, hard and often discouraging. Success largely depends on work, self-confidence and effort rather than inspiration. Thomas Alva Edison expressed this idea claiming that 'Genius is one percent inspiration, ninety-nine percent perspiration' and the validity of this statement can be illustrated by analyzing the creative process that led James Clerk Maxwell to the derivation of the electromagnetic synthesis.

The first contribution to this field dates back to 1855, when Maxwell published '*On Faraday's lines of force*', in which a set of coupled differential equations relating electricity and magnetism was put forward. At this point, it should be mentioned that Maxwell used to face different problems simultaneously but letting long periods of time pass between successive publications about the same matter. The advantage of this way of proceeding, which is usual among physicists, is that ideas can be thought out and analyzed from different points of view. Accordingly, the next work on the electromagnetic theory, '*On physical lines of force*', was published in 1861. From 1855 to 1861, his efforts were devoted to a motley assortment of issues including color analysis, the rings of Saturn or the kinetic theory of gases. In

1861, Maxwell found that the speed of propagation of an electromagnetic field is approximately that of the speed of light. This finding made him return to his research in electromagnetism. Working on the problem further, Maxwell predicted the existence of waves of oscillating electric and magnetic fields that travel through empty space at the speed of light. But this did not happen like greased lightning. The theory was firstly published in two preliminary works in 1865 and 1868, and fully developed in 'A Treatise on Electricity and Magnetism', in 1873. It was a laborious task, which took eighteen years of intense intellectual work.

Finally, it is worth mentioning that Maxwell does not fit the stereotype of physicist described above in spite of being one of the most outstanding physicists. Maxwell was a rich man, owner of a 2000-acre ranch, and expert swimmer and rider. In addition, he had a great sense of humor, as many of his letters reveal, and some talent for poetry. As a poet, he was usually cheerful but he also wrote some profound poems. With the aim of contributing to a comprehensive image of Maxwell, we end this essay with the last verse of a poem written by this great man to his wife:

All powers of mind, all force of will,
May lie in dust when we are dead,
But love is ours, and shall be still,
When earth and seas are fled.

Maxwell in 1855, when he was 24 years old.

SOLVED PROBLEMS

Problem 1.1. Imagine again a long straight wire (whose radius is negligible) carrying a steady current I along the z-axis (in the $+z$-direction, see Figure 1.4). Show that Gauss's law for magnetism holds at any point $P(x,y,z)$ outside the wire. Note that we had already shown that this field satisfies Ampère-Maxwell's law in section 1.3.3.

Solution. As discussed previously, the Cartesian components of the magnetic field created by an infinitely long straight wire (whose radius is negligible) carrying a steady current I along the z-axis are given by:

$$B_x = -B \sin \varphi = -\frac{\mu_0 I}{2\pi r}\frac{y}{r} = -\frac{\mu_0 I y}{2\pi r^2}$$

$$B_y = B \cos \varphi = \frac{\mu_0 I}{2\pi r}\frac{x}{r} = \frac{\mu_0 I x}{2\pi r^2}$$

$$B_z = 0$$

Gauss's law for magnetism states that:

$$\vec{\nabla} \cdot \vec{B} = \frac{\partial B_x}{\partial x} + \frac{\partial B_y}{\partial y} + \frac{\partial B_z}{\partial z} = 0$$

We must therefore take partial derivatives recalling that $r^2 = x^2 + y^2$. In general, you may need to compute partial derivatives of r^n, where $r = \sqrt{x^2 + y^2 + z^2}$ or $r = \sqrt{x^2 + y^2}$, dealing with problems that exhibit spherical or cylindrical symmetry. This can be done applying the chain rule for the composition of two or more functions as follows:

$$\frac{\partial r^n}{\partial x} = \frac{\partial r^n}{\partial r}\frac{\partial r}{\partial x} = nr^{n-1} \cdot \frac{2x}{2\sqrt{x^2 + y^2}} = nr^{n-1} \cdot \frac{x}{r} = nxr^{n-2}$$

Applying this result yields:

$$\frac{\partial B_x}{\partial x} = \frac{\partial}{\partial x}\left(-\frac{\mu_0 I y}{2\pi r^2}\right) = -\frac{\mu_0 I y}{2\pi}\frac{\partial r^{-2}}{\partial x} = \frac{\mu_0 I y x}{\pi r^4}$$

$$\frac{\partial B_y}{\partial y} = \frac{\partial}{\partial y}\left(\frac{\mu_0 I x}{2\pi r^2}\right) = \frac{\mu_0 I x}{2\pi r^2}\frac{\partial r^{-2}}{\partial y} = -\frac{\mu_0 I x y}{\pi r^4}$$

$$\frac{\partial B_z}{\partial z} = 0$$

Adding these partial derivatives gives:

$$\vec{\nabla}\cdot\vec{B} = \frac{\partial B_x}{\partial x} + \frac{\partial B_y}{\partial y} + \frac{\partial B_z}{\partial z} = \frac{\mu_0 I y x}{\pi r^3} - \frac{\mu_0 I y x}{\pi r^3} = 0$$

Thus this field satisfies Gauss's law.

Problem 1.2. Now consider the magnetic field of a long straight wire, whose radius is R. This wire carries a current I that is uniformly distributed over its circular cross-section. The magnetic field inside the wire has also circular field lines but its magnitude is given by:

$$B = \frac{\mu_0 I r}{2\pi R^2} \quad (r \le R)$$

This result can be derived from the integral form of Ampère's law. Here you must however verify that the differential forms of Ampère's law and Gauss's law are also satisfied.

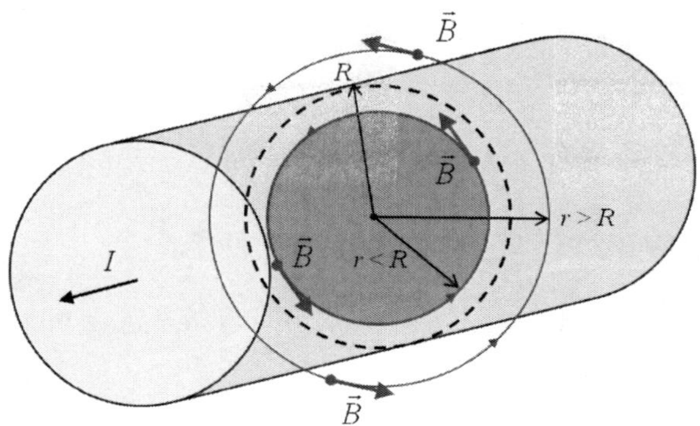

Solution. Let us begin with Gauss's law. As you can guess, first the Cartesian components of the field must be expressed as a function of x, y and z, and functions of these variables, such as r, which is the magnitude of the vector position. As the field lines are circular, you can follow a method similar to that used in section 1.3.4 (see Figure 1.4 and Equations 1.51), which in this case gives:

$$B_x = -\frac{\mu_0 Ir}{2\pi R^2}\frac{y}{r} = -\frac{\mu_0 Iy}{2\pi R^2}$$

$$B_y = \frac{\mu_0 Ir}{2\pi R^2}\frac{x}{r} = \frac{\mu_0 Ix}{2\pi R^2}$$

$$B_z = 0$$

Note that B_x does not depend on x and B_y does not depend on y. Consequently:

$$\frac{\partial B_x}{\partial x} = 0$$

$$\frac{\partial B_y}{\partial y} = 0$$

$$\frac{\partial B_z}{\partial z} = 0$$

Obviusly, the addition of these partial derivatives yields:

$$\vec{\nabla} \cdot \vec{B} = \frac{\partial B_x}{\partial x} + \frac{\partial B_y}{\partial y} + \frac{\partial B_z}{\partial z} = 0$$

Thus the validity of Gauss's law is confirmed.

Now, let us consider Ampère-Maxwell' law recalling that its differential form states:

$$\vec{\nabla} \times \vec{B} = \mu_0 \left(\vec{J} + \varepsilon_0 \frac{\partial \vec{E}}{\partial t} \right)$$

First, we will compute the curl of the magnetic field:

$$\vec{\nabla} \times \vec{B} = \begin{vmatrix} \hat{i} & \hat{j} & \hat{k} \\ \frac{\partial}{\partial x} & \frac{\partial}{\partial y} & 0 \\ -\frac{\mu_0 I y}{2\pi R^2} & \frac{\mu_0 I x}{2\pi R^2} & 0 \end{vmatrix} =$$

$$= \frac{\partial}{\partial x}\left(\frac{\mu_0 I x}{2\pi R^2} \right)\hat{k} - \frac{\partial}{\partial y}\left(-\frac{\mu_0 I y}{2\pi R^2} \right)\hat{k} =$$

$$= \frac{\mu_0 I}{2\pi R^2}\hat{k} + \frac{\mu_0 I}{2\pi R^2}\hat{k} = \frac{\mu_0 I}{\pi R^2}\hat{k}$$

On the right side we find the local sources of the magnetic field. First note that there is no electric field anywhere, so:

$$\frac{\partial \vec{E}}{\partial t} = 0$$

The current density must also be computed, which can be done recalling that the current is the flux of the current density:

$$I = \int_S \vec{J} \cdot d\vec{A}$$

In this case, both $\vec{J}$ and $d\vec{A}$ have the $+z$-direction. In addition, $\vec{J}$ has the same value over S, so it can come out of the integral. Thus $J = I / \pi R^2$ and $\vec{J} = I / \pi R^2 \hat{k}$. The vector obtained multiplying this current density by the magnetic permeability of vacuum is identical to the curl of the magnetic field, which proves the validity of Ampère-Maxwell' law.

Problem 1.3. An inventor is looking for financial support for the construction of a perpetual motion machine of the first kind that would considerably reduce the price of electricity if it worked. According to his project, which is available for investors, the cornerstone of the machine is a coil that creates a magnetic field. If the coil is oriented in the x-direction, then the magnetic field must be given by $\vec{B}(x) = ax^2 \hat{i}$, where x is the distance to the center of the coil and a is a constant. Find out if this magnetic field is feasible verifying Ampère's law and Gauss's law for it. In view of the result, would you recommend this project to investors?

Solution. The differential form of Ampère-Maxwell's law states that:

$$\vec{\nabla} \times \vec{B} = \mu_0 \left(\vec{J} + \varepsilon_0 \frac{\partial \vec{E}}{\partial t} \right)$$

There are not currents or electric fields at any point of the inner space of the coil. Thus $\vec{J} = 0$ and $\dfrac{\partial \vec{E}}{\partial t} = 0$ .Consequently, the curl of the magnetic field must be zero. This can be confirmed computing the curl as:

$$\vec{\nabla} \times \vec{B} \equiv \begin{vmatrix} \hat{i} & \hat{j} & \hat{k} \\ \frac{\partial}{\partial x} & 0 & 0 \\ ax^2 & 0 & 0 \end{vmatrix} = 0$$

We therefore conclude that this magnetic field does verify Ampère-Maxwell's law. What happens with Gauss's law? Let us answer to this question computing the divergence of the field:

$$\vec{\nabla} \cdot \vec{B} = \frac{\partial(ax^2)}{\partial x} = 2ax$$

However, according to Gauss's law, the divergence of the magnetic field must be zero everywhere:

$$\vec{\nabla} \cdot \vec{B} = \frac{\partial B_x}{\partial x} + \frac{\partial B_y}{\partial y} + \frac{\partial B_z}{\partial z} = 0$$

Thus we conclude that neither this magnetic field nor the machine can exist. We should not recommend this project!

Problem 1.4. Consider again the electric field created by a point charge (see section 2.4). Given that this field is electrostatic (not generated by a time-varying magnetic field), it should verify that $\vec{\nabla} \times \vec{E} = 0$ everywhere except the point where the charge is placed. Prove it explicitly.

Solution. The electric field of point charge located in the origin of the coordinate system is given by:

$$\vec{E} = K \frac{q}{r^2} \hat{r} = K \frac{q}{r^3} \vec{r}$$

Its Cartesian components are:

$$E_x = K \frac{qx}{r^3}$$

$$E_y = K \frac{qy}{r^3}$$

$$E_z = K \frac{qz}{r^3}$$

Do not forget that r is function of x, y and z. First, let us compute the curl of this field:

$$\vec{\nabla} \times \vec{E} = \begin{vmatrix} \hat{i} & \hat{j} & \hat{k} \\ \dfrac{\partial}{\partial x} & \dfrac{\partial}{\partial y} & \dfrac{\partial}{\partial z} \\ \dfrac{Kqx}{r^3} & \dfrac{Kqy}{r^3} & \dfrac{Kqz}{r^3} \end{vmatrix} =$$

$$= \frac{\partial}{\partial y}\left(K\frac{qz}{r^3}\right)\hat{i} + \frac{\partial}{\partial x}\left(K\frac{qy}{r^3}\right)\hat{k} +$$

$$+ \frac{\partial}{\partial z}\left(K\frac{qy}{r^3}\right)\hat{j} - \frac{\partial}{\partial y}\left(K\frac{qx}{r^3}\right)\hat{k}$$

$$- \frac{\partial}{\partial x}\left(K\frac{qz}{r^3}\right)\hat{j} - \frac{\partial}{\partial z}\left(K\frac{qy}{r^3}\right)\hat{i} =$$

$$= \left[Kqz\frac{\partial r^{-3}}{\partial y} - Kqy\frac{\partial r^{-3}}{\partial z}\right]\hat{i} +$$

$$+ \left[Kqx\frac{\partial r^{-3}}{\partial z} - Kqz\frac{\partial r^{-3}}{\partial x}\right]\hat{j} +$$

$$+ \left[Kqy\frac{\partial r^{-3}}{\partial x} - Kqx\frac{\partial r^{-3}}{\partial y}\right]\hat{k} =$$

Recalling that $\partial r^n / \partial x = nxr^{n-2}$ (see Problem 1.1) gives:

$$\vec{\nabla} \times \vec{E} = \left[\frac{-3Kqzy}{r^5} + \frac{3Kqzy}{r^5}\right]\hat{i}$$

$$+ \left[\frac{-3Kqzx}{r^5} + \frac{3Kqzx}{r^5}\right]\hat{j} +$$

$$+\left[\frac{-3Kqyx}{r^5}+\frac{3Kqyx}{r^5}\right]\hat{k}=$$

$$=0\hat{i}+0\hat{j}+0\hat{k}=0$$

The curl is zero everywhere (except at the origin), as expected.

Problem 1.5. Consider again a uniformly charged sphere and the electric field created *inside* it (see section 2.5). Given that this field is electrostatic (not generated by a time-varying magnetic field), it should verify that $\vec{\nabla}\times\vec{E}=0$. Prove it explicitly.

Solution. The electric field created inside this sphere (whose charge and radius are q and a, respectively) is given by:

$$\vec{E}=K\frac{q}{a^3}\vec{r}$$

Being its Cartesian components:

$$E_x=K\frac{qx}{a^3}$$

$$E_y=K\frac{qy}{a^3}$$

$$E_z=K\frac{qz}{a^3}$$

The curl of this field can be computed expanding the determinant:

$$\vec{\nabla}\times\vec{E}\equiv\begin{vmatrix}\hat{i} & \hat{j} & \hat{k}\\[4pt] \dfrac{\partial}{\partial x} & \dfrac{\partial}{\partial y} & \dfrac{\partial}{\partial z}\\[8pt] K\dfrac{qx}{a^3} & K\dfrac{qy}{a^3} & K\dfrac{qz}{a^3}\end{vmatrix}=$$

$$= \frac{\partial}{\partial y}\left(K\frac{qz}{a^3} \right)\hat{i} + \frac{\partial}{\partial z}\left(K\frac{qx}{a^3} \right)\hat{j}$$

$$+ \frac{\partial}{\partial x}\left(K\frac{qy}{a^3} \right)\hat{k} - \frac{\partial}{\partial y}\left(K\frac{qx}{a^3} \right)\hat{k}$$

$$- \frac{\partial}{\partial z}\left(K\frac{qy}{a^3} \right)\hat{i} - \frac{\partial}{\partial x}\left(K\frac{qz}{a^3} \right)\hat{j} = 0$$

This curl is zero because all partial derivatives are zero.

Problem 1.6. The electric field created by charges can always be derived from an electrostatic potential, $V(\vec{r})$:

$$\vec{E} = -\vec{\nabla}V = -\frac{\partial V}{\partial x}\hat{i} - \frac{\partial V}{\partial y}\hat{j} - \frac{\partial V}{\partial z}\hat{k}$$

From Gauss's law, show that the electrostatic potential must satisfy Poisson's equation:

$$\frac{\partial^2 V}{\partial x^2} + \frac{\partial^2 V}{\partial y^2} + \frac{\partial^2 V}{\partial z^2} = -\frac{\rho}{\varepsilon_0}$$

At points of space where there is not charge, the charge density is zero and the resulting equation is then known as Laplace's equation:

$$\frac{\partial^2 V}{\partial x^2} + \frac{\partial^2 V}{\partial y^2} + \frac{\partial^2 V}{\partial z^2} = 0$$

Solution. Gauss's law in differential form states:

$$\vec{\nabla} \cdot \vec{E} = \frac{\rho}{\varepsilon_o}$$

Let us compute the divergence expressing the electric field as the gradient of the electrostatic potential:

$$\vec{\nabla} \cdot \vec{E} = \left(\frac{\partial}{\partial x}\hat{i} + \frac{\partial}{\partial y}\hat{j} + \frac{\partial}{\partial z}\hat{k} \right)\left(-\frac{\partial V}{\partial x}\hat{i} - \frac{\partial V}{\partial y}\hat{j} - \frac{\partial V}{\partial z}\hat{k} \right)$$

$$= -\frac{\partial}{\partial x}\left(\frac{\partial V}{\partial x} \right) - \frac{\partial}{\partial y}\left(\frac{\partial V}{\partial y} \right) - \frac{\partial}{\partial z}\left(\frac{\partial V}{\partial z} \right) =$$

$$= -\frac{\partial^2 V}{\partial x^2} - \frac{\partial^2 V}{\partial y^2} - \frac{\partial^2 V}{\partial z^2}$$

Such a straightforward calculation leads to Poisson's equation.

Problem 1.7. Imagine a fluid that is rotating counterclockwise like a rigid body around the z-axis with angular velocity ω. Show that the curl of the corresponding velocity field is $\vec{\nabla} \times \vec{v} = 2\omega\hat{k}$.

Note: The curl of the velocity vector is usually known as vorticity in fluid dynamics.

Solution. First, you must calculate the Cartesian components of the velocity field. As the field lines are circular, you can draw a figure similar to Figure 1.4 but replacing the magnetic field vector by the velocity vector. From this figure, it can be shown that the velocity field of this fluid is given by:

$$v_x = -v\sin\varphi = -\omega r\frac{y}{r} = -\omega y$$

$$v_y = v\cos\varphi = \omega r\frac{x}{r} = \omega x$$

$$v_z = 0$$

Here, v is the magnitude of the velocity vector and φ the angle that the position vector of an arbitrary point of space forms with the x-axis (see Figure 1.4). From these components, the curl is computed as follows:

$$\vec{\nabla} \times \vec{v} = \begin{vmatrix} \hat{i} & \hat{j} & \hat{k} \\ \frac{\partial}{\partial x} & \frac{\partial}{\partial y} & 0 \\ -\omega y & \omega x & 0 \end{vmatrix} =$$

$$= \frac{\partial}{\partial x}(\omega x)\hat{k} - \frac{\partial}{\partial y}(-\omega y)\hat{k} =$$

$$= \omega\hat{k} + \omega\hat{k} = 2\omega\hat{k}$$

As you can see, this curl has a nonzero and constant value everywhere. This means that a tiny wheel paddle would rotate with the same angular velocity (ω) at any point of space. This seems logical at the origin of the coordinate system because the fluid rotates around the z-axis with this velocity but maybe you are wondering why the wheel paddle would rotate at any other point. This can be justified as follows. The magnitude of the velocity at a given point of space is proportional to the distance to the rotation axis, $v = \omega r$. Thus the velocity at the points of the wheel closer to the z-axis is smaller than the velocity at the points more distant from it, which would make the wheel rotate.

Go further. Now imagine that the angular velocity of the fluid decreases with the distance to the z-axis according to $\omega = c/r$, where c is a constant. Consequently, the magnitude of the velocity of any element of the fluid is constant, $v = \omega r = c$ and the wheel paddle would not rotate in this case. You can prove it explicitly computing the curl.

It is not knowledge, but the act of learning, not possession but the act of getting there, which grants the greatest enjoyment.

C. F. Gauss.

Chapter 2

ELECTROMAGNETIC WAVES IN FREE SPACE

2.1. PLANE ELECTROMAGNETIC WAVES

2.1.1. Maxwell's Equations in the Absence of Electric Charges and Currents

In Chapter 1, the differential form of Maxwell's equations was derived starting from the integral form. In this chapter we will learn how Maxwell's equations lead to the wave equation without requiring complex identities of the classical theory of fields. In other words, the differential formalism will be applied to the free propagation of electromagnetic waves in empty space (in which no charges or currents exist). Let us begin thinking about the physical meaning and mathematical appearance of Maxwell's equations:

$$\vec{\nabla} \cdot \vec{E} = \rho / \varepsilon_0$$

$$\vec{\nabla} \cdot \vec{B} = 0$$

$$\vec{\nabla} \times \vec{E} = -\frac{\partial \vec{B}}{\partial t} \qquad (2.1)$$

$$\vec{\nabla} \times \vec{B} = \mu_0 \left(\vec{J} + \varepsilon_0 \frac{\partial \vec{E}}{\partial t} \right)$$

These equations tell us that the electric and magnetic fields have different sources and we might then expect that their behaviors are different as well. For example, the divergence of the magnetic field is always zero and its field lines cannot initiate or terminate at any point of space. However, electric fields have field lines that originate and terminate on positive and negative charges respectively and its divergence is non-zero at those points. Now let us examine the right side of Faraday's law and Ampère-Maxwell's law. According to these equations the electric field has only one vector source (changing magnetic fields), but the magnetic field has two kinds of vector sources (electric currents and changing electric fields). Thus the electric and magnetic field seem to exhibit certain asymmetry.

Let us now consider an electromagnetic wave traveling through empty space. There cannot be charges or currents at any point of this medium. Then Maxwell's equations adopt a simplified form:

$$\vec{\nabla} \cdot \vec{E} = 0$$

$$\vec{\nabla} \cdot \vec{B} = 0$$

$$\vec{\nabla} \times \vec{E} = -\frac{\partial \vec{B}}{\partial t} \tag{2.2}$$

$$\vec{\nabla} \times \vec{B} = \mu_0 \varepsilon_0 \frac{\partial \vec{E}}{\partial t}$$

If multiplicative constants (such as the minus sign in Faraday's law or $\mu_0 \varepsilon_0$) are ignored, certain symmetry comes out in these equations:

1) Both fields have zero-divergence. Their field lines have no points of initiation or termination.
2) Time-varying magnetic fields are sources of electric field (Faraday's law) and time-varying electric fields are sources of magnetic fields (Ampère-Maxwell's law).

Thus, both fields behave similarly. This symmetry will be deeply analyzed in the rest of the chapter. At this point it should be stressed that Faraday's law and Ampère-Maxwell's law help us understand how an electromagnetic wave propagates: An oscillating magnetic field at a given point induces an oscillating electric field in its neighborhood, which in turn generates a changing magnetic field and so on. If the mechanism of propagation is qualitatively understood, the matter can be mathematically addressed. In this journey, we will find the answer to this question: What speed do electromagnetic waves travel?

2.1.2. Transverse Character of Electromagnetic Waves

To begin with, let us consider a plane electromagnetic wave traveling in a given direction. Without loss of generality, we will assume that it is $+x$-direction, which makes calculations easier because our functions will not depend on y or z. Initially, we should calculate three Cartesian components for the electric field and three Cartesian components for the magnetic field:

$$E_x(x,t), E_y(x,t), E_z(x,t)$$
$$B_x(x,t), B_y(x,t), B_z(x,t) \tag{2.3}$$

Applying Gauss's law for the electric field gives:

$$\vec{\nabla} \cdot \vec{E} = \frac{\partial E_x}{\partial x} + \frac{\partial E_y}{\partial y} + \frac{\partial E_z}{\partial z} = 0 \tag{2.4}$$

Given that none of these components depends on y or z, two of these derivatives are zero:

$$\frac{\partial E_x}{\partial x} + 0 + 0 = 0 \tag{2.5}$$

If this partial derivative is zero, the x-component of the electric field does not depend on x. Obviously, Gauss's law for magnetism leads to a similar conclusion for the x-component of the magnetic field. In other words, the x-components of both fields might initially depend on t, $E_x(t)$, $B_x(t)$. However, we will rigorously prove that the x-component of both fields does not depend on time either.

Ampère-Maxwell's law will be now applied to these fields (in the absence of currents):

$$\vec{\nabla} \times \vec{B} = \mu_0 \varepsilon_0 \frac{\partial \vec{E}}{\partial t} \tag{2.6}$$

Taking into account that $\partial/\partial y = 0$, $\partial/\partial z = 0$ because none of the components depends on either y or z, the two sides of this law are:

$$\vec{\nabla} \times \vec{B} \equiv \begin{vmatrix} \hat{i} & \hat{j} & \hat{k} \\ \frac{\partial}{\partial x} & 0 & 0 \\ B_x & B_y & B_z \end{vmatrix} = -\frac{\partial B_z}{\partial x}\hat{j} + \frac{\partial B_y}{\partial x}\hat{k} \tag{2.7}$$

$$\mu_0 \varepsilon_0 \frac{\partial \vec{E}}{\partial t} = \mu_0 \varepsilon_0 \left(\frac{\partial E_x}{\partial t}\hat{i} + \frac{\partial E_y}{\partial t}\hat{j} + \frac{\partial E_z}{\partial t}\hat{k} \right) \tag{2.8}$$

Thus, Ampère-Maxwell's law can be written as:

$$-\frac{\partial B_z}{\partial x}\hat{j} + \frac{\partial B_y}{\partial x}\hat{k} = \mu_0 \varepsilon_0 \left(\frac{\partial E_x}{\partial t}\hat{i} + \frac{\partial E_y}{\partial t}\hat{j} + \frac{\partial E_z}{\partial t}\hat{k} \right) \tag{2.9}$$

This vector equation is equivalent to three scalar equations:

$$\frac{\partial E_x}{\partial t} = 0 \tag{2.10}$$

$$-\frac{\partial B_z}{\partial x} = \mu_0 \varepsilon_0 \frac{\partial E_y}{\partial t} \tag{2.11}$$

$$\frac{\partial B_y}{\partial x} = \mu_0 \varepsilon_0 \frac{\partial E_z}{\partial t} \tag{2.12}$$

The first of these three equations absolutely proves that the x-component of the electric field does not depend on t either, as mentioned before. This component should therefore be constant. However, we are interested in wave solutions, which simultaneously depend on space and time. Thus, for our purposes, we might assume $E_x = 0$ without loss of generality. In short, the electric field must be perpendicular to the direction of propagation.

Let us imagine that this direction remains constant. Then, one of the coordinate axes (let us say the y-axis) can be oriented pointing in this well-defined direction. In this case we say that the electric field is linearly polarized in the y-direction and the electric field has only y-component, so $E_z = 0$. According to Equation 2.12:

$$\frac{\partial B_y}{\partial x} = 0 \tag{2.13}$$

Now let us apply Faraday's law considering the preceding conclusions. The curl of the electric field can be written as:

$$\vec{\nabla} \times \vec{E} \equiv \begin{vmatrix} \hat{i} & \hat{j} & \hat{k} \\ \frac{\partial}{\partial x} & 0 & 0 \\ 0 & E_y & 0 \end{vmatrix} = \frac{\partial E_y}{\partial x} \hat{k} \tag{2.14}$$

Thus Faraday's law leads to these three scalar equations:

$$0 = -\frac{\partial B_x}{\partial t} \tag{2.15}$$

$$0 = -\frac{\partial B_y}{\partial t} \tag{2.16}$$

$$\frac{\partial E_y}{\partial x} = -\frac{\partial B_z}{\partial t} \tag{2.17}$$

From the first of these three equations we conclude that the x-component of the magnetic field cannot depend on t, as mentioned before. As done in the case of the electric field, we can assume that $B_x = 0$ if we are interested in wave solutions. Thus we finally conclude that both fields are perpendicular to the direction of propagation, which means we are dealing with a transverse wave, because the longitudinal components equal zero.

Equations 2.13 and 2.16 prove that the y-component of the magnetic field does not depend on x or t either. If constant values are not properly considered wave solutions, we can also assume that $B_y = 0$. The reader should note that the magnetic field lies along the z-axis if the electric field lies in the y-axis. In other words, the magnetic field is perpendicular to both the electric field and the direction of propagation. Figure 2.1 illustrates the relative orientation of these two fields.

Thus applying Maxwell's equations we have concluded that only two components of the initial set of unknown quantities (Expression 2.2) are required. If the wave propagates in the x-direction and the electric field is linearly polarized in y-direction, these components are E_y and B_z. The electric and magnetic fields are therefore perpendicular to each other and to the propagation direction. These two fields are additionally related by Equations 2.11 and 2.17, which are versions of Ampère-Maxwell's and Faraday's laws particularized for the situation we are analyzing.

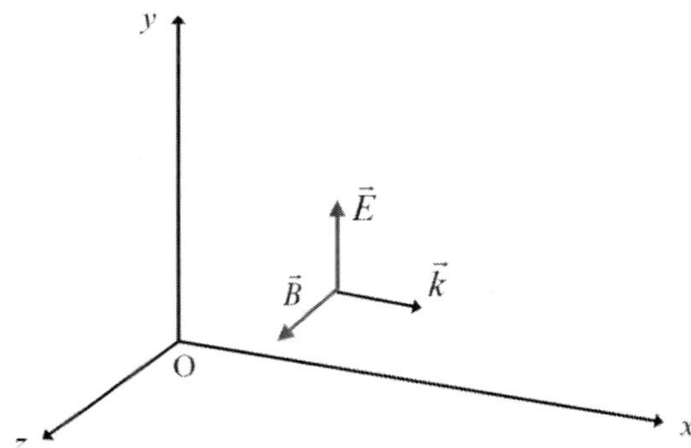

Figure 2.1. Relative orientation of the electric and magnetic fields and the direction of propagation for an electromagnetic wave traveling in the +x-direction and linearly polarized in the y-direction.

2.1.3. Derivation of the Wave Equation

Let us retrieve the simplified version of Ampère-Maxwell's law (Equation 2.11). B_z can be eliminated from this equation by differentiating both sides with respect to time:

$$\frac{\partial}{\partial t}\left(\frac{\partial B_z}{\partial x}\right) = -\frac{\partial}{\partial t}\left(\mu_0\varepsilon_0\frac{\partial E_y}{\partial t}\right) \tag{2.18}$$

On the left side, the z-component of the magnetic field is first differentiated with respect to x and then respect to t. If a function of two variables has continuous second partial derivatives at any given point, Schwarz's theorem states that the order of taking partial derivatives can be interchanged:

$$\frac{\partial}{\partial t}\left(\frac{\partial B_z}{\partial x}\right) = \frac{\partial}{\partial x}\left(\frac{\partial B_z}{\partial t}\right) \tag{2.19}$$

On the other hand, $\mu_0\varepsilon_0$ can come out of time derivative on the right side of Equation 2.18, which leads to a second derivative of E_y:

$$\frac{\partial}{\partial x}\left(\frac{\partial B_z}{\partial t}\right) = -\mu_0\varepsilon_0\frac{\partial^2 E_y}{\partial t^2} \tag{2.20}$$

Inserting Faraday's law (Equation 2.17) into the left side of this Equation 2.20 yields:

$$\frac{\partial}{\partial x}\left(-\frac{\partial E_y}{\partial x}\right) = -\mu_0\varepsilon_0\frac{\partial^2 E_y}{\partial t^2} \tag{2.21}$$

By simplifying we finally obtain:

$$\frac{\partial^2 E_y}{\partial x^2} = \mu_0\varepsilon_0\frac{\partial^2 E_y}{\partial t^2} \tag{2.22}$$

It should be mentioned that this equation only involves one unknown function: the y-component of the electric field. What happens with the z-component of the magnetic field? A similar result can be obtained for B_z if both sides of the reduced version of Faraday's law are differentiated with respect to t, the order of partial derivatives is again interchanged and, finally, Ampère-Maxwell's law (Equation 2.11) is inserted to eliminate E_y:

$$\frac{\partial^2 B_z}{\partial x^2} = \mu_0 \varepsilon_0 \frac{\partial^2 B_z}{\partial t^2} \tag{2.23}$$

Equations 2.22 and 2.23 are the result we were looking for. Mathematically speaking, both are identical to the differential equation that waves on a string obey:

$$\frac{\partial^2 y}{\partial x^2} = \frac{1}{v^2} \frac{\partial^2 y}{\partial t^2} \tag{2.24}$$

where v is the velocity of propagation of the wave and y is the displacement of the string (with respect to equilibrium). In our case, the disturbance traveling through empty space is the electromagnetic field itself. In addition, the comparison between Equations 2.24 and 2.23 (or 2.22) allows us to compute the velocity of propagation of this electromagnetic disturbance:

$$v = \frac{1}{\sqrt{\varepsilon_0 \mu_0}} \tag{2.25}$$

Inserting into this equation the values of the permittivity and permeability of empty space gives 300000 km/s, in round numbers, which is just the speed of light in free space.

Maxwell found this fascinating coincidence in the second half of the 19th century, as previously commented in Chapter 1. He showed that electromagnetic disturbances obey a differential wave equation. This was not a minor finding. Needless to say that nowadays it is not easy to imagine how our everyday life would be without electromagnetic waves. What is more, we are worried about electromagnetic pollution. However, there was not experimental evidence of electromagnetic waves when Maxwell theoretically proved their existence by deriving Equations 2.22 and 2.23. Electromagnetic waves were first produced and detected in laboratory by Heinrich Hertz some years later (in 1887), which meant the beginning of the age of telecommunications.

In addition, Maxwell concluded that light was an electromagnetic wave. This was not a trivial finding either. From Newton's times, the nature of light was a controversial matter. This famous physicist supported that light was composed of corpuscles. One of his arguments against the wave nature of light was that waves were known to bend around obstacles, while light traveled in straight lines.

On the other hand, Christian Huygens argued in favor of the wave nature of light and, more than a century later, Thomas Young showed that light behaved as waves by means of a diffraction experiment. The weakness of the wave theory was that light, like sound, would presumably need a medium for transmission.

Thus the existence of a hypothetical substance pervading all bodies, called *luminiferous aether*, was postulated. The ether was a really magical substance. It had to be lighter than air. Otherwise the orbit of planets would be visibly affected. However, it should be more rigid than steel because the speed of light was much higher than the speed of sound in steel. These troubles disappear keeping in mind that light is not a mechanical wave but an electromagnetic wave, which can propagate through empty space.

2.1.4. Harmonic Plane Waves

It is well known that the solution to the differential wave equation must have the functional form $f(t \mp x/v)$, where the minus sign is used for waves traveling in the $+x$-direction whereas the plus sign is used for the propagation in the $-x$-direction.

Recalling that $k = \omega/v$, this solution can be written in the form $f(\omega t \mp kx)$. Here, we will restrict ourselves to the case in which f is a harmonic function (sine or cosine). Fourier analysis states that any other solution can be expressed as a superposition of harmonic functions.

First, let us examine the case of a wave propagating in the $+x$-direction. If the electric and magnetic fields are described by harmonic solutions, we can write:

$$E_y(x,t) = E_0 \sin(\omega t - kx + \varphi)$$
$$B_z(x,t) = B_0 \sin(\omega' t - k' x + \varphi')$$

$$(2.26)$$

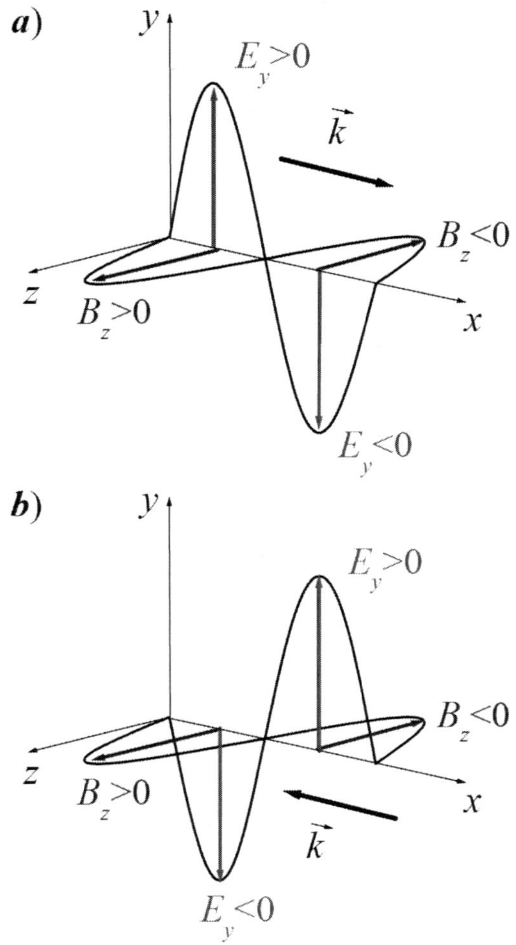

Figure 2.2. Oscillations of the electric and magnetic fields for an electromagnetic wave linearly polarized in the *y*-direction and traveling: a) in the +*x*-direction; b) in the −*x*-direction.

You should note that, in principle, we have considered the possibility that these two fields have different frequencies, wavenumbers and phase constants. Equations 2.22 and 2.23 tell us that both have the same velocity of propagation but, *a priori*, there is not anything suggesting that the rest of parameters must be equal for both.

However, these fields must satisfy Maxwell's equations and, particularly, Faraday's law, which for this situation is:

$$\frac{\partial E_y}{\partial x} = -\frac{\partial B_z}{\partial t} \qquad (2.27)$$

Inserting the harmonic solutions into Faraday's law gives:

$$-kE_0 \cos(\omega t - kx + \varphi) = -\omega' B_0 \cos(\omega' t - k'x + \varphi') \qquad (2.28)$$

This equality of two harmonic functions must be satisfied for any t-value and x-value, and this can only happen if the arguments of both functions are identical. Thus:

$$\omega = \omega', k = k', \varphi = \varphi' \qquad (2.29)$$

The amplitudes of these harmonic functions must be identical as well:

$$E_0 = cB_0 \qquad (2.30)$$

We therefore conclude that the frequencies, wavenumbers and phase constant must be the same for both fields, and their amplitudes are proportional. The proportionality constant in empty space is c, the speed of light. This also means that both fields are in phase, reaching their maxima together. In fact, the proportionality between them takes place at any time and point of space:

$$E_y(x,t) = cB_z(x,t) \qquad (2.31)$$

In summary, these fields behave in the same manner, which agrees with the symmetry observed for the electric and magnetic field in Maxwell's equations when waves propagate through empty space.

Now let analyze the vector product $\vec{E} \times \vec{B}$. For the wave described previously this product is proportional to the unit vector $\hat{j} \times \hat{k} = \hat{i}$, which points in the direction of propagation.

This situation is illustrated in Figure 2.2. What happens when the wave propagates in the $-x$-direction (see also Figure 2.2)? In this case the harmonic solutions have the form:

$$E_y(x,t) = E_0\sin(\omega t + kx + \varphi)$$
$$B_z(x,t) = B_0\sin(\omega't + k'x + \varphi')$$
$$(2.32)$$

By following the same line of reasoning, it can be easily proved that:

$$\omega = \omega', k = k', \varphi = \varphi' \pm \pi \tag{2.33}$$

$$E_0 = cB_0 \tag{2.34}$$

As can be seen, the only difference from the previous case is that the phase constants differ in π radians and:

$$E_y(x,t) = E_0\sin(\omega t + kx + \varphi)$$
$$B_z(x,t) = B_0\sin(\omega t + kx + \varphi \pm \pi) = -B_0\sin(\omega t + kx + \varphi)$$
$$(2.35)$$

Unlike the propagation in the +x-direction, the fields are now in phase opposition (π radians out of phase). When one of them reaches the maximum value, the other passes though the minimum value:

$$E_y(x,t) = -cB_z(x,t) \tag{2.36}$$

However, $\vec{E} \times \vec{B}$ lies again in the direction of propagation, because this vector product is now proportional to $-\hat{i}$. In fact, this holds in general: $\vec{E} \times \vec{B}$ always points in the direction of propagation, whether the wave propagates in the +x-direction or any other.

2.2. ENERGY OF ELECTROMAGNETIC WAVES

2.2.1. Energy Density

The establishment of an electric field in a region of space (for example, between the plates of a capacitor) requires some work, which can be thought of as energy stored in the points of space where the electric field exists. In this case, it is possible to define the energy density (at a given point) as the energy per unit of volume. It can be shown that the local energy density due to the existence of an electric field is given by:

$$u_E = \frac{1}{2}\varepsilon_0 E^2 \tag{2.37}$$

On the other hand, the establishment of a magnetic field in a region of space where it did not exist before (for example around a coil) also requires some work. In this case, we can think that the corresponding energy is stored in the points where the magnetic field already exists. The local energy density due to this magnetic field is given by:

$$u_B = \frac{1}{2\mu_0} B^2 \tag{2.38}$$

An electromagnetic wave creates electric and magnetic fields at the points of space where it exists. Consequently, the total energy density is the sum of the electric and magnetic energy densities:

$$u(\vec{r},t) = \frac{1}{2}\varepsilon_0 E^2(\vec{r},t) + \frac{1}{2\mu_0} B^2(\vec{r},t) \tag{2.39}$$

This expression can be transformed keeping in mind that $E(\vec{r},t) = \pm cB(\vec{r},t)$ (see Equations 2.31 and 2.36):

$$u(\vec{r},t) = \frac{1}{2}\varepsilon_0 E^2 + \frac{1}{2\mu_0}(E/c)^2 \tag{2.40}$$

Recalling that $c^2 = 1/(\varepsilon_0\mu_0)$ it is easy to conclude that:

$$u(\vec{r},t) = \frac{1}{2}\varepsilon_0 E^2 + \frac{1}{2}\varepsilon_0 E^2 = \varepsilon_0 E^2(\vec{r},t) \tag{2.41}$$

This discussion also proves that the energy densities associated the electric and magnetic fields of an electromagnetic wave have the same value. As can be seen, the energy densities associated to the electric and magnetic fields also exhibit symmetry.

2.2.2. Energy Flux of an Electromagnetic Wave

Let us consider again an electromagnetic wave propagating in the $+x$-direction and also an imaginary rectangular prism that has infinitesimal dimensions dx, dy and dz. This prism is so thick that the fields hardly change along dx. Under such circumstances, the energy stored (dU) in this prism can be calculated from the energy density at this region of space (given by Equation 2.41) multiplied by its volume:

$$dU = u(x,t)dx \cdot dy \cdot dz = \varepsilon_0 E^2 dx \cdot dy \cdot dz \qquad (2.42)$$

But waves propagate and the energy stored in this prism travels with the wave. In the time it takes the wave to travel the distance dx (which we will refer to as dt), the electromagnetic energy dU passes through one the sides perpendicular to the direction of propagation. The area of these sides is $dy{\cdot}dz$. Taking into account that $dt = dx/c$, the flux of energy per unit time and unit area is then given by:

$$S(x,t) = \frac{dU}{dy \cdot dz \cdot dt} = \frac{udx \cdot dy \cdot dz}{dy \cdot dz \cdot dx/c} = cu(x,t) = c\varepsilon_0 E^2 \qquad (2.43)$$

At any point in the region where an electromagnetic wave exists we can state that $E(x,t) = cB(x,t)$. In addition, $c^2 = 1/(\varepsilon_0 \mu_0)$. With the help of the previous results we can write:

$$S(x,t) = \frac{EB}{\mu_0} \qquad (2.44)$$

As the electric and magnetic fields are perpendicular to each other, Equation 2.44 can be rewritten in the form:

$$S(x,t) = \frac{\left|\vec{E} \times \vec{B}\right|}{\mu_0} \tag{2.45}$$

From Equation 2.45, the *Poynting vector* can be defined as:

$$\vec{S} = \frac{\vec{E} \times \vec{B}}{\mu_0} \tag{2.46}$$

According to Equation 2.44, the magnitude of this vector gives the instantaneous flux of energy per unit time and area whereas its direction points in the direction of propagation of the wave (see end of section 1.5). This property has been inferred for a plane wave, but its validity is much more general.

2.2.3. Intensity of a Harmonic Electromagnetic Wave

In any case, it should be stressed that the Poynting vector previously defined varies with time. For harmonic waves of high frequency, the instantaneous Poynting vector oscillates so rapidly that it would be hardly monitored. Then, it is quite useful to assess its time-averaged value, which is known as intensity. The definition of intensity as the time-averaged power per unit area can be applied to any wave, regardless of its nature. For a harmonic electromagnetic wave:

$$I = \langle S(x,t) \rangle = \langle \varepsilon_0 c E^2 \rangle = \langle \varepsilon_0 c E_0^2 \sin^2(\omega t - kx) \rangle \tag{2.47}$$

where the brackets $\langle\ \rangle$ stands for time averaging. Although this operator has a formal definition, we will here apply a simple and intuitive reasoning to calculate the time average appearing in Equation 2.47. The function whose time average we want to compute oscillates between 0 (when the sine is zero) and $\varepsilon_0 c E_0^2$ (when the sine is ± 1. Since this oscillation is completely regular and well defined, we can easily admit that the average is given by the arithmetic mean of these two values:

$$I = \frac{1}{2}\varepsilon_0 c E_0^2 \tag{2.48}$$

HISTORICAL NOTE: ELECTROMAGNETIC THEORY, OLIVER HEAVISIDE AND THE INVENTION OF VECTOR ANALYSIS

Physics and mathematics have often evolved hand in hand, particularly since the publication of '*Philosophiæ Naturalis Principia Mathematica*' by Newton. Concerning the foundations of classical mechanics, the value of this work is undeniable. However, it also provided guidelines for the method used in the generation of knowledge by natural philosophy first and physics later: the scientific method. This way of proceeding has been shown to be extremely efficient unraveling the secrets of universe and its efficiency improves when mathematical descriptions and procedures are preferred rather than the merely qualitative ones. In any case, the development of classical physics during its golden age (the 18th and 19th centuries) required the creation of new mathematical tools for the accurate description of reality. Let us discuss, just as an example, the case of the electromagnetic theory, vector analysis and Oliver Heaviside.

As pointed out in a previous note, Maxwell published his book '*A Treatise on Electricity and Magnetism*' in 1873 but, initially, this book did not have a big influence on physics for different reasons. On the one hand, the mathematical description was extremely complicated and only a few contemporary scientists could understand it. The notion of displacement current was not well understood either. On the other hand, there was not experimental evidence of the existence of the displacement current or electromagnetic waves.

Oliver Heaviside was one of the few scientists of the same epoch that studied Maxwell's work in depth, being very impressed with its content despite the fact that the mathematical terminology made it difficult to understand. It took several years a thorough study of Maxwell's treatise and he began to quote it in his own works in 1876. The early death of Maxwell when he was only forty eight meant a radical change of circumstances. The master would not make contributions to his theory any longer. In addition, the theory should be spread among the scientific community in a more understandable way.

Heaviside took this task on and began to perform it in 1882, according to his own account. It should be stressed that he did not restrict himself to a mere repetition of the work. Heaviside reformulated, refined and extended Maxwell's theory. Nowadays, we commonly say that the electromagnetic theory was synthesized by Maxwell in four equations, but this statement is not strictly true. In 1864, Maxwell summarized the behavior of electric and magnetic fields in twenty equations involving twenty variables ('*A Dynamical Theory of the Electromagnetic Field*', Chapter III) in which he had included Ohm's law, the continuity equation for charge, and the constitutive relation between the electric field and the electric displacement. Leaving out these laws, which today are not included in the set we refer to as Maxwell's equations, his electromagnetic synthesis would be mathematically described in thirteen equations, which were modified, simplified and expressed in the form we know by Heaviside and Hertz (they did it independently at the beginning). For sake of simplicity, the scalar and vector potentials were eliminated from the original formulation but, even with these changes, the resulting set of equations would not resemble the current one due to the mathematical representation. In relation to this point, it should be mentioned that physical properties involving direction had been described by Maxwell in terms of quaternions, a mathematical tool introduced by the Irish mathematician William Rowan Hamilton some years before. However, quaternions were not suitable for pedagogic purposes. Thus Heaviside developed vector analysis, which was introduced in the form we currently know in Chapter III of '*Electromagnetic theory*'. What is more, he argued against quaternions, getting into heated arguments with P.G. Tait, their main supporter. As vector analysis was conceptually simpler and notationally cleaner, quaternions were eventually replaced by vectors in most of the branches of physics.

SOLVED PROBLEMS

Problem 2.1. The light of a laser beam propagating through empty space in the +z-direction can be described by a harmonic electromagnetic wave whose frequency is $8 \cdot 10^{13}$ Hz. The magnetic field lies in the y-direction and its amplitude is 4·10-4 T. Write the equations describing the electric and magnetic fields.

Solution. From the wording we have $\vec{B}(z,t) = B_y(z,t)\,\hat{j}$. Recalling that $\vec{E} \times \vec{B}$ gives the direction of propagation and $\hat{i} \times \hat{j} = \hat{k}$ we can conclude that:

$$\vec{E}(z,t) = E_x(z,t)\hat{i}$$

Being an harmonic wave we can specifically write:

$$\vec{E}(z,t) = E_0 \sin(\omega t - kz)\hat{i}$$
$$\vec{B}(z,t) = B_0 \sin(\omega t - kz)\hat{j}$$

Here, the wave parameters are given by:

$$\omega = 2\pi f = 2\pi \cdot 8 \cdot 10^{13} = 5.03 \cdot 10^{14} \; rad/s$$

$$k = \frac{\omega}{c} = \frac{5.03 \cdot 10^{14}}{3 \cdot 10^{8}} = 1.68 \cdot 10^{6} \; m^{-1}$$

$$E_0 = cB_0 = 3 \cdot 10^{8}\, 4 \cdot 10^{-4} = 1.2 \cdot 10^{5}\, N/C$$

Consequently, the equations describing the behavior of the electric and magnetic fields in time and space are:

$$\vec{E}(z,t) = 1.2 \cdot 10^{5} \sin(5.03 \cdot 10^{14} t - 1.68 \cdot 10^{6} z)\hat{i}$$
$$\vec{B}(z,t) = 4 \cdot 10^{-4} \sin(5.03 \cdot 10^{14} t - 1.68 \cdot 10^{6} z)\hat{j}$$
$$\text{(SI units)}$$

> **Problem 2.2.** The electric field of an electromagnetic wave propagating through empty space is given by:
>
> $$\vec{E}(y,t) = 2.0\cdot 10^5 \sin(1.5\cdot 10^{14} t - ky)\,\hat{k}$$
>
> All the parameters are expressed in SI units. What is its wavelength? Find out the wave equation for the magnetic field.

Solution. From the data provided in the wording, we straightfowrdrly conclude that $\omega = 1.5\cdot 10^{14}$ rad/s. Other parameters are:

$$k = \frac{\omega}{c} = \frac{1.5\cdot 10^{14}}{3\cdot 10^8} = 5\cdot 10^5 \; m^{-1}$$

$$\lambda = \frac{2\pi}{k} = \frac{2\pi}{5\cdot 10^5} = 1.26\cdot 10^{-5}\, m$$

As the wave propagates in the $+y$-direction and $\vec{E}\times\vec{B}$ points in this direction, the magnetic field must point in the x-direction (since $\hat{k}\times\hat{i}=\hat{j}$):

$$\vec{B}(y,t) = B_0 \sin(\omega t - ky)\hat{i}$$

The amplitude of the magnetic field can be easily computed from:

$$B_0 = \frac{E_0}{c} = \frac{2.0\cdot 10^5}{3\cdot 10^8} = 6.67\cdot 10^{-4}\, T$$

Inserting the values calculated previously in the expression of $\vec{B}(y,t)$ gives:

$$\vec{B}(y,t) = 6.67\cdot 10^{-4} \sin(1.5\cdot 10^{14} t - 5\cdot 10^5 y)\hat{i} \quad \text{(SI units)}$$

Problem 2.3. The electric and magnetic fields of an electromagnetic wave are given by:

$$\vec{E} = 30\cos(9x + 27 \cdot 10^8 t)\hat{j}$$

$$\vec{B} = -10^{-7}\cos(9x + 27 \cdot 10^8 t)\hat{k}$$

The amplitudes of the electric and magnetic fields are expressed in N/C and T, respectively. The rest of parameters are also expressed in SI units. What is the frequency of this wave? What is the wavelength? What is the speed of propagation? Prove that this wave verifies Faraday's law and Ampère-Maxwell's law.

Solution. From the expressions of the field we straightforwardly deduce that $\omega = 27 \cdot 10^8$ rad/s and $k = 9$ m^{-1}. Other parameters are easily calculated from them:

$$f = \frac{\omega}{2\pi} = \frac{27 \cdot 10^8}{2\pi} = 4.30 \cdot 10^8 \ Hz$$

$$\lambda = \frac{2\pi}{k} = \frac{2\pi}{9} = 0.70\,m$$

The direction of progation is given by $\vec{E} \times \vec{B}$, which points in the $-x$-direction. This can also be inferred from the arguments of the cosine functions. The speed of propagation can be computed as the quotient between the amplitudes of the electric and magnetic fields:

$$v = E_0 / B_0 = 3 \cdot 10^8 \ m/s$$

Now, let us verify two Maxwell's equations. Faraday's law states that:

$$\vec{\nabla} \times \vec{E} = -\frac{\partial \vec{B}}{\partial t}$$

First, the curl of the electric field will be calculated by expanding the determinant:

$$\vec{\nabla}\times\vec{E} \equiv \begin{vmatrix} \hat{i} & \hat{j} & \hat{k} \\ \frac{\partial}{\partial x} & 0 & 0 \\ 0 & E_y(x,t) & 0 \end{vmatrix} = \frac{\partial E_y(x,t)}{\partial x}\hat{k} =$$

$$-30{\cdot}9\cdot\sin(9x+27{\cdot}10^8\,t)\,\hat{k} =$$

$$-270\cdot\sin(9x+27{\cdot}10^8\,t)\,\hat{k} \quad N/(C{\cdot}m)$$

On the other hand, the right side gives:

$$-\frac{\partial\vec{B}}{\partial t} = -(-1)\cdot(-10^{-7})\cdot27{\cdot}10^8\sin(9x+27{\cdot}10^8\,t)\,\hat{k}$$

$$= -270\sin(9x+27{\cdot}10^8\,t)\,\hat{k} \quad T/s$$

As can be seen, both results are identical, as expected. According to Ampère-Maxwell's law:

$$\vec{\nabla}\times\vec{B} = \mu_0\left(\vec{J}+\varepsilon_0\frac{\partial\vec{E}}{\partial t}\right)$$

In empty space, there are not currents, $\vec{J}=0$. Thus:

$$\vec{\nabla}\times\vec{B} = \mu_0\varepsilon_0\frac{\partial\vec{E}}{\partial t}$$

Computing the curl yields:

$$\vec{\nabla}\times\vec{B} \equiv \begin{vmatrix} \hat{i} & \hat{j} & \hat{k} \\ \frac{\partial}{\partial x} & 0 & 0 \\ 0 & 0 & B_z(x,t) \end{vmatrix} = -\frac{\partial B_z(x,t)}{\partial x}\hat{j} =$$

$$= -(-1)\cdot(-1)10^{-7}\cdot9\sin(9x+27{\cdot}10^8\,t)\,\hat{j} =$$

$$= -9{\cdot}10^{-7}\sin(9x+27{\cdot}10^8\,t)\,\hat{j} \quad T/m$$

Whereas the right side gives:

$$\mu_0\varepsilon_0\frac{\partial\vec{E}}{\partial t}=-4\pi\cdot10^{-7}\cdot8.85\cdot10^{-12}\cdot27\cdot10^8\cdot30\cdot\sin(9x+27\cdot10^8 t)\,\hat{j}=$$

$$=-9\cdot10^{-7}\sin(9x+27\cdot10^8 t)\,\hat{j}\quad T/m$$

Both results are again identical, which confirms that this wave satisfies Ampère-Maxwell's law as well.

Problem 2.4. The Poynting vector of an electromagnetic waves propagating through empty space is given by:

$$\vec{S}(x,t)=10^4\cos^2(10x-3\cdot10^9 t)\,\hat{i}\ W/m^2$$

where x and t are expressed in SI units. Its electric field is parallel to the y-axis. What is the direction of propagation? What is the direction of the magnetic field? Determine the wavelength, the frequency and the intensity of this wave. Finally, write the expressions of the electric and magnetic fields.

Solution. The Poyinting vector tells us the direction of propagation of an electromagnetic wave. Thus this wave propagates in a direction parallel to the x-axis.

The direction of the magnetic field can be find out recalling that the Poyiting vector is given by:

$$\vec{S}=\frac{\vec{E}\times\vec{B}}{\mu_0}$$

According to the wording of this problem, $\vec{S}=S\hat{i}$ and $\vec{E}=E\hat{j}$. Consequently, the magnetic field must be parallel to the z-axis, $\vec{B}=B\hat{k}$, because $\hat{j}\times\hat{k}=\hat{i}$.

From the expression of the Poying vector we can also deduce that ω is $3 \cdot 10^9$ rad/s and the wavenumber is 10 m^{-1}. From them, the frequency and the wavelength are straightforwardly worked out:

$$f = \frac{\omega}{2\pi} = \frac{3 \cdot 10^9}{2\pi} = 4.77 \cdot 10^8 \ Hz$$

$$\lambda = \frac{2\pi}{k} = \frac{2\pi}{10} = 0.63 m$$

The intensity of an electromagnetic wave is the time-averaged magnitude of the Poynting vector, as discussed previously. Recalling that the time average of the square of harmonic functions is ½ yields:

$$I = \langle S(x,t) \rangle = \langle 10^4 \cos^2(10x - 3 \cdot 10^9 t) \rangle = 5000 \ W/m^2$$

This value is also useful for the calculation of the amplitude of the electric field because:

$$I = \frac{1}{2} \varepsilon_0 c E_0^2$$

Solving for the amplitude of the electric field gives:

$$E_0 = \left(\frac{2I}{\varepsilon_0 c} \right)^{1/2} = \left(\frac{2 \cdot 5000}{8.85 \cdot 10^{-12} \cdot 3 \cdot 10^8} \right)^{1/2} = 1941 \ N/C$$

In addition, the amplitudes of both fields are proportional:

$$E_0 = c B_0$$

Thus:

$$B_0 = \frac{E_0}{c} = \frac{1941}{3 \cdot 10^8} = 6.47 \cdot 10^{-6} \ T$$

Having calculated all these data, the vector expressions of the fields turn out to be:

$$\vec{E}(x,t) = 1941\cos(10x - 3{\cdot}10^9 t)\,\hat{j} \quad N/C$$

$$\vec{B}(x,t) = 6.47{\cdot}10^{-6}\cos(10x - 3{\cdot}10^9 t)\,\hat{k} \quad T$$

Problem 2.5. An electromagnetic wave propagates in the $-z$-direction with a frequency of 5 MHz. Its magnetic field is parallel to the x-axis and has an amplitude of $70{\cdot}10^{-6}$ T.

a) Find the vector expressions of the associated fields taking into account that the magnetic field takes its maximum value for $z=0$ and $t=0$.

b) Find the Poyiting vector and its time average.

c) Check that the fields that you have worked out verify Maxwell's equations.

Solution.

a) Let first calculate the angular frequency, the wavelength and the wavenumber:

$$\omega = 2\pi f = 2\pi{\cdot}5{\cdot}10^6 = 3.14{\cdot}10^7 \; rad/s$$

$$\lambda = \frac{c}{f} = \frac{3{\cdot}10^8}{5{\cdot}10^6} = 60\,m$$

$$k = \frac{2\pi}{\lambda} = \frac{2\pi}{60} = 0.105\,m^{-1}$$

From these data and those included in the wording, we can compute the magnetic field:

$$\vec{B}(z,t) = 70{\cdot}10^{-6}\cos(3.14{\cdot}10^7 t + 0.105z)\,\hat{i} \quad T$$

Note that we have written a cosine function since it takes a maximum value at $z=0$ and $t=0$. If we had chosen the sine function, the field would vanish under such conditions.

The amplitude of the electric field can be easily computed as follows:

$$E_0 = cB_0 = 3 \cdot 10^8 \cdot 70 \cdot 10^{-6} = 2.10 \cdot 10^4 \, N/C$$

The direction of the magnetic field can be find out recalling that the Poyiting vector is given by:

$$\vec{S} = \frac{\vec{E} \times \vec{B}}{\mu_0}$$

According to the wording of this problem, $\vec{S} = -S\hat{k}$ and $\vec{B} = B\hat{i}$. Consequently, the electric field must be parallel to the x-axis, $\vec{E} = E\hat{i}$, because $\hat{j} \times \hat{i} = -\hat{k}$. Thus:

$$\vec{E}(z,t) = 2.10 \cdot 10^4 \cos(3.14 \cdot 10^7 t + 0.105z)\,\hat{j}$$

b) The Poynting vector is computed from the fields as:

$$\vec{S}(z,t) = \frac{\vec{E}(z,t) \times \vec{B}(z,t)}{\mu_0} = -1.17 \cdot 10^6 \cdot \cos^2(3.14 \cdot 10^7 t + 0.105z)\,\hat{k}$$

The time average of its magnitude (which gives us the intensity) is evaluated recalling that the time average of the square of harmonic functions is ½:

$$I = \langle S(z,t) \rangle = 5.85 \cdot 10^5 \, W/m^2$$

c) Let us begin with the verification of Gauss's law in differential form:

$$\vec{\nabla} \cdot \vec{E} = \frac{\partial E_x}{\partial x} + \frac{\partial E_y}{\partial y} + \frac{\partial E_z}{\partial z} = \frac{\rho}{\varepsilon_o}$$

In empty space the charge density is zero everywhere and:

$$\vec{\nabla} \cdot \vec{E} = \frac{\partial E_x}{\partial x} + \frac{\partial E_y}{\partial y} + \frac{\partial E_z}{\partial z} = 0$$

In this case the field only has y-component, $\vec{E}(z,t) = E_y(z,t)\hat{j}$, and can be easily shown that:

$$\frac{\partial E_y(z,t)}{\partial y} = 0$$

Gauss's law for magnetism can be verified in a similar way. As the magnetic field is parallel to the x-axis, $\vec{B}(z,t) = B_x(z,t)\hat{i}$, only one of the three partial derivatives is involved. But this partial derivative is zero:

$$\frac{\partial B_x(z,t)}{\partial x} = 0$$

Consequently:

$$\vec{\nabla} \cdot \vec{B} = \frac{\partial B_x}{\partial x} + \frac{\partial B_y}{\partial y} + \frac{\partial B_z}{\partial z} = 0$$

At this point, let us focus on Faraday's law:

$$\vec{\nabla} \times \vec{E} = -\frac{\partial \vec{B}}{\partial t}$$

To begin with, the curl is computed:

$$\vec{\nabla}\times\vec{E} \equiv \begin{vmatrix} \hat{i} & \hat{j} & \hat{k} \\ 0 & 0 & \dfrac{\partial}{\partial z} \\ 0 & E_y(z,t) & 0 \end{vmatrix} = -\dfrac{\partial E_y(z,t)}{\partial z}\hat{i} =$$

$$=(-1)\cdot(-1)\cdot 2.10\cdot 10^4 \cdot 0.105\cdot \sin(3.14\cdot 10^7\, t+0.105z)\hat{i} =$$

$$=2.20\cdot 10^3\cdot \sin(3.14\cdot 10^7\, t+0.105z)\hat{i} \quad N/(C\cdot m)$$

On the other hand, the right side gives:

$$-\frac{\partial \vec{B}}{\partial t}=-(-1)\cdot 70\cdot 10^{-6}\cdot 3.14\cdot 10^7\cdot \sin(3.14\cdot 10^7\, t+0.105z)\hat{i} =$$

$$=2.20\cdot 10^3\cdot \sin(3.14\cdot 10^7\, t+0.105z)\hat{i} \quad T/s$$

Both results are identical. Thus these fields obey Faraday's law. Finally, let us verify Ampère-Maxwell's law:

$$\vec{\nabla}\times\vec{B} = \mu_0\left(\vec{J}+\varepsilon_0\frac{\partial \vec{E}}{\partial t}\right)$$

In the absence of currents, this law becomes: $\vec{\nabla}\times\vec{B} = \mu_0\varepsilon_0\dfrac{\partial \vec{E}}{\partial t}$

First, we will compute the curl of the magnetic field:

$$\vec{\nabla}\times\vec{B} \equiv \begin{vmatrix} \hat{i} & \hat{j} & \hat{k} \\ 0 & 0 & \dfrac{\partial}{\partial z} \\ B_x(z,t) & 0 & 0 \end{vmatrix} = -\dfrac{\partial B_x(z,t)}{\partial z}\hat{j} =$$

$$= (-1)\cdot 70\cdot 10^{-6}\cdot 0.105\cdot \sin(3.14\cdot 10^7\, t+0.105z)\,\hat{j} =$$

$$=-7.3\cdot 10^{-6}\cdot \sin(3.14\cdot 10^7\, t+0.105z)\,\hat{j} \quad T/m$$

The right side turns out to be:

$$\mu_0\varepsilon_0\frac{\partial \vec{E}}{\partial t} =$$

$$-4\pi{\cdot}10^{-7}{\cdot}8.85{\cdot}10^{-12}{\cdot}2.10{\cdot}10^4{\cdot}3.14{\cdot}10^7\,\sin(3.14{\cdot}10^7 t + 0.105z)\,\hat{j} =$$

$$= -7.3{\cdot}10^{-6}{\cdot}\sin(3.14{\cdot}10^7 t + 0.105z)\,\hat{j}$$

Again both sides agree (rounding off), which proves that these fields also satisfy Ampère-Maxwell's law.

Problem 2.6. Consider again the linearly polarized electromagnetic wave employed in section 2.1. It has been shown that the electric field satisfies a differential wave equation. Now prove that the magnetic field also obeys this equation:

$$\frac{\partial^2 B_z}{\partial x^2} = \mu_0\varepsilon_0\frac{\partial^2 B_z}{\partial t^2}$$

Solution. Our starting point is Faraday's law particularized for an electromagnetic wave whose electric field is parallel to the y-axis (Equation 2.17):

$$\frac{\partial E_y}{\partial x} = -\frac{\partial B_z}{\partial t}$$

E_y can be eliminated from this equation as follows. First, both sides are differentiated with respect to time:

$$\frac{\partial}{\partial t}\left(\frac{\partial E_y}{\partial x}\right) = \frac{\partial}{\partial t}\left(-\frac{\partial B_z}{\partial t}\right)$$

Obviously, the right side involves the second derivative of B_z with respect to time:

$$\frac{\partial}{\partial t}\left(\frac{\partial E_y}{\partial x}\right) = -\frac{\partial^2 B_z}{\partial t^2}$$

The order of the partial derivatives on the left side is interchanged assuming that Schwarz's theorem is valid:

$$\frac{\partial}{\partial x}\left(\frac{\partial E_y}{\partial t}\right) = -\frac{\partial^2 B_z}{\partial t^2}$$

Recall the version of Ampère-Maxwell's law particularized for this case (Equation 2.11):

$$\frac{\partial E_y}{\partial t} = -\frac{1}{\mu_0 \varepsilon_0}\frac{\partial B_z}{\partial x}$$

If this result is inserted into the previous equation, we have:

$$\frac{\partial}{\partial x}\left(-\frac{1}{\mu_0 \varepsilon_0}\frac{\partial B_z}{\partial x}\right) = -\frac{\partial^2 B_z}{\partial t^2}$$

After simplifying, we end up with:

$$\frac{1}{\mu_0 \varepsilon_0}\frac{\partial^2 B_z}{\partial x^2} = \frac{\partial^2 B_z}{\partial t^2}$$

We live in a society exquisitely dependent on science and technology, in which hardly anyone knows anything about science and technology.

Carl Sagan, in *The Skeptical Inquirer.*

Chapter 3

GUIDED ELECTROMAGNETIC WAVES

3.1. INTRODUCTION

Chapter 2 was devoted to the propagation of plane electromagnetic waves in free space. The intensity of these waves remains constant while traveling. However, it should be kept in mind that they are an idealization that allows us to simplify the mathematical treatment of propagation but, in practice, plane wavefronts are hardly ever observed. In general, the surface of the wavefront grows when the wave moves away from its source. Thus intensity decreases with distance (a noticeable exception is a laser beam). For example, in the case of spherical wavefronts, intensity decreases with the square of the distance. In any case, repeaters must be used to increase the range of a transmitted signal by retransmission. Only in this way, signals propagating through free space can cover long distances.

The propagation in free space presents another important drawback that can sometimes make it little efficient: obstacles in the way. Electromagnetic waves can go through dielectric materials but, when they find an obstacle, a considerable fraction of the incident energy may be reflected. Consequently, obstacles can weaken electromagnetic waves to a great extent and even block its way. They can also bring about unwanted interference phenomena.

These two shortcomings of the propagation in open space can be overcome confining waves inside the so-called waveguides. These structures restrict the propagation to one dimension (in the guiding direction), so that the energy is not scattered in different directions. In addition, waves can find a way around obstacles. For sound waves, a tube can work as a waveguide. In the case of electromagnetic waves, there are different structures that allow guided propagation. However, the original and most common one is a hollow metal pipe.

This chapter does not attempt to perform a rigorous, extensive and in-depth study of electromagnetic waveguides, which can be found in many excellent textbooks of advanced electromagnetism. Our aim is to present two examples of waveguides that can be analyzed with relatively simple mathematical tools and illustrate the differences between free and guided propagation. The first example is the parallel-plate waveguide. We can imagine that electromagnetic waves travel along the guide in a zigzag path, being repeatedly reflected between two opposite conducting walls. Again, this is an idealized picture to simplify the mathematical treatment, but we can find similar real situations. In fact, the first transatlantic transmission of radio signals from England to Canada is an epoch-making event in the history of telecommunications that can be explained in these terms (see historical note of this chapter for further details). The propagation in a dielectric waveguide may be viewed in the same way, with the waves confined inside the dielectric by total internal reflection. Our second example is a hollow metal tube with a rectangular cross-section. Rectangular waveguides are used routinely to transfer large amounts of microwave power at frequencies greater than 3 GHz. In any case, the propagation along this waveguide can also be described as the process resulting from successive reflections at the conducting walls. This picture will allow us to derive expressions for the fields or the cutoff frequencies without previous knowledge of the theory of functions of a complex variable or partial differential equations.

3.2. The Parallel-Plate Waveguide: TE Modes

3.2.1. A Boundary Condition for the Electric Field

Before facing the problem of propagation, we will show that the tangential component of the electric field must vanish at the surface of a perfectly conducting wall. Inside a conducting medium, the electric field and the current density are related by Ohm's law in local form, $\vec{E} = \rho\vec{J}$, where ρ is the resistivity.

A perfect conductor is an idealized material whose resistivity tends to zero. Thus, the electric field inside a *perfectly* conducting medium has to vanish provided that the current density is finite.

But what happens in the immediate vicinity of this material? The integral form of Faraday's law can help us to find the answer to this question. Let us apply this law to a very small rectangle with two sides parallel to the boundary between the perfectly conducting medium and empty space, as shown in Figure 3.1.

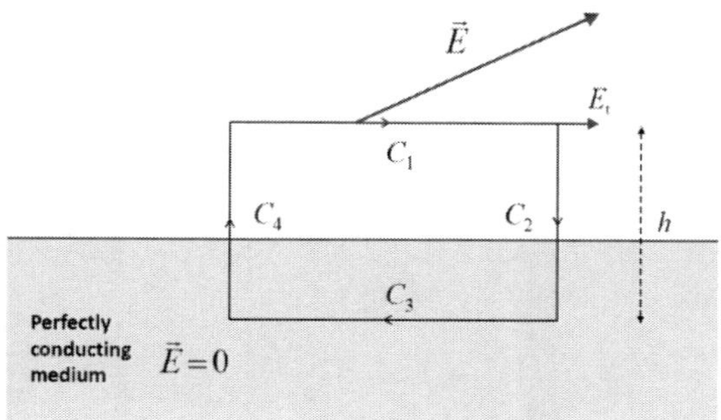

Figure 3.1. Circulation around a small rectangle crossed by the surface of a perfectly conducting wall.

The circulation appearing on the left side of Faraday's law can be split into four line integrals, corresponding to the four sides of the rectangle:

$$\oint_C \vec{E}\cdot d\vec{l} = \int_{C1}\vec{E}\cdot d\vec{l} + \int_{C2}\vec{E}\cdot d\vec{l} + \int_{C3}\vec{E}\cdot d\vec{l} + \int_{C4}\vec{E}\cdot d\vec{l} = -\frac{d\Phi_B}{dt} \qquad (3.1)$$

where C1, C2, C3 and C4 stand for these four sides. The integral along C3 is zero because the electric field is zero inside the perfect conductor, as stated above. Now imagine the limiting case in which the sides perpendicular to the boundary shrink to zero ($h \to 0$). In this limit, the magnetic flux through the rectangle vanishes because its area tends to zero. Moreover, the line integrals along C2 and C4 also vanish when $h \to 0$. Consequently, the line integral along C1 must also equal zero when we approach the boundary. This will happen if the tangential component of the electric field is zero at the surface.

3.2.2. Electric Field in TE Modes

Let us analyze the propagation of electromagnetic waves along a parallel-plate guide. Figure 3.2 illustrates the situation: two perfectly conducting plates placed at $x = 0$ and $x = a$, so the separation between them is a. Now imagine a wave traveling in a direction forming an angle θ with the y-axis. When this wave hits a perfectly conducting plate it is reflected (like in a mirror), changing its direction of propagation until it strikes the opposite plate, and so on. Figure 3.2 shows a couple of rays and their corresponding reflections.

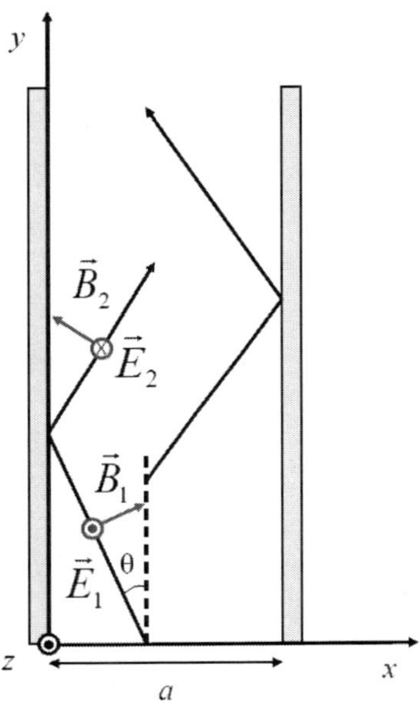

Figure 3.2. Successive reflections in a parallel-plate waveguide. The wave resulting from the superposition of wave 1 and wave 2 propagates along the +y-direction. Note that the electric field has no component in the direction of propagation (TE mode).

We have drawn only two rays, but you should keep in mind that the waves corresponding to both directions fill the whole guide. In other words, in any point inside the guide there will be a superposition of two waves traveling in two different directions. One of the waves comes from the plate on the right (wave 1) whereas the other goes to this plate after reflection (wave 2). They form angles θ (counterclockwise) and -θ (clockwise) with the y-direction, respectively. We will show that their interference travels along the +y-direction, which will be referred to as the guiding direction.

To begin with, let us consider the case in which the electric field of these two waves has no component in the guiding direction (y-component according to our coordinate system). This propagation mode is known as *transverse electric* (TE). As discussed in the previous chapter, the electric fields of these two waves must be perpendicular to their propagation directions. In addition, they must be perpendicular to the y-direction. Thus the electric field must be parallel to the z-axis and only have z-component. Obviously, the z-component of the field resulting from the superposition can be obtained adding the z-components of both waves:

$$E_z(\vec{r},t) = E_{1z}(\vec{r},t) + E_{2z}(\vec{r},t) \tag{3.2}$$

where $\vec{r}$ is the position vector and t stands for time, as usual. At this point, the reader should bring the equation for two-dimensional plane waves to mind. An appendix has been included at the end of this chapter for this purpose. Assuming harmonic behavior, the equations for these two waves are:

$$E_{1z}(\vec{r},t) = E_{10}\sin(\omega t - k\hat{u}_1 \cdot \vec{r} + \phi_1)$$
$$E_{2z}(\vec{r},t) = E_{20}\sin(\omega t - k\hat{u}_2 \cdot \vec{r} + \phi_2) \tag{3.3}$$

where E_{10} y E_{20} are their amplitudes, ω is the angular frequency, k is the wavenumber, ϕ_1 and ϕ_2 are the phase constants, and $\hat{u}_1$ and $\hat{u}_2$ are unit vectors pointing in the corresponding directions of propagation. It can be easily shown that the unit vectors for waves 1 and 2 are given by:

$$\hat{u}_1 = -\sin\theta\,\hat{i} + \cos\theta\,\hat{j}$$
$$\hat{u}_2 = \sin\theta\,\hat{i} + \cos\theta\,\hat{j} \tag{3.4}$$

Inserting Equation 3.4 into Equation 3.3 and expanding scalar products yields:

$$E_{1z} = E_{10}\sin(\omega t + kx\sin\theta - ky\cos\theta + \phi_1)$$
$$E_{2z} = E_{20}\sin(\omega t - kx\sin\theta - ky\cos\theta + \phi_2) \tag{3.5}$$

At first sight, we could think that amplitudes and phase constants can take arbitrary values but, at this point, we should bring to mind that the tangential component of the electric field must be zero at the surface of a perfectly conduction wall. In our case, the field only has z-component, which is tangential to the walls located at $x = 0$ and $x = a$. Thus:

$$E_z(0, y, t) = 0$$
$$E_z(a, y, t) = 0 \tag{3.6}$$

These are boundary conditions for the electric field at the perfectly conducting surfaces bounding the guide. It should be pointed out that such boundary conditions are analogous to those imposed to a wave on a string fixed at both ends, $y(0,t) = 0$, $y(L,t) = 0$, where y is the displacement describing the wave and L is the length of the string. This analogy will help us understand subsequent results. First, let us apply the boundary condition at $x = 0$, which gives:

$$E_{10}\sin(\omega t - ky\cos\theta + \phi_1) + E_{20}\sin(\omega t - ky\cos\theta + \phi_2) = 0 \tag{3.7}$$

Thus we have a sum of two harmonic functions of the same frequency and wavenumber that must be zero for any y- and t-value. This will happen if they have the same amplitude and are out of phase π radians:

$$E_{10} = E_{20} \equiv E_0$$
$$\phi_2 - \phi_1 = \pi \tag{3.8}$$

where E_0 is the amplitude of both waves. Thus the fields that must be added are:

$$E_{1z} = E_0\sin(\omega t + kx\sin\theta - ky\cos\theta + \phi_1)$$
$$E_{2z} = E_0\sin(\omega t - kx\sin\theta - ky\cos\theta + \phi_1 + \pi) \tag{3.9}$$

After factoring E_0, the addition of two sine functions must be computed. This task can be carried out recalling a useful trigonometric identity:

$$\sin\alpha + \sin\beta = 2\sin\left(\frac{\alpha+\beta}{2}\right)\cos\left(\frac{\alpha-\beta}{2}\right) \tag{3.10}$$

Indentifying α and β with the arguments of the two sine functions and applying the trigonometric identity yields:

$$E_z = 2E_0 \sin(\omega t - ky\cos\theta + \tfrac{\phi_1 + \phi_2}{2})\cos(-kx\,sen\theta + \pi/2) \qquad (3.11)$$

The mean value of both phase constants can be denoted as $\phi = (\phi_1 + \phi_2)/2$. Taking into account that $\cos(\pi/2 - \delta) = \sin\delta$ (the sine and cosine of complementary angles are identical) we finally obtain:

$$E_z = 2E_0 \sin(kx\sin\theta)\sin(\omega t - ky\cos\theta + \phi) \qquad (3.12)$$

Before discussing this important result, Equation 3.12 can be written in a more compact form defining the following concepts:

$$k_g \equiv k\cos\theta$$
$$k_c \equiv k\sin\theta \qquad (3.13)$$

k_g is the propagation wave number along the guide direction (the subscript g stands for *guide*), also known as propagation constant. In many textbooks, this concept is also denoted by β. The second expression defines the so-called cutoff wave number. In terms of these quantities, the transverse electric field is given by:

$$E_z(x, y, t) = 2E_0 \sin(k_c x)\sin(\omega t - k_g y + \phi) \qquad (3.14)$$

As can be seen, this field depends on both space variables, x and y. This is logical because waves 1 and 2, whose superposition produces this field, also depend on them. However, Equation 3.14 reveals that the behaviors in the x- and y-directions are quite different. Concerning the vertical direction, the y-variable is coupled to time in the usual form for traveling waves, $\omega t - ky + \phi$, although the wave number in empty space has been replaced by the propagation wave number along the guide direction, k_g. Obviously, this also means that the wavelength in the guiding direction differs from the wavelength in free space. As the frequency remains the same, the velocity of propagation in the guide must also differ from the velocity in free space. This noticeable discrepancy will be discussed in depth later.

Now, let us analyze the behavior in the horizontal direction (x-variable). At a glance, we can see that this space variable is not coupled to time, which is a typical feature of standing waves. We therefore find that the field behaves as a traveling wave in the y-direction and as a standing wave in the x-direction. The standing behavior in the horizontal direction is a result of the confinement between the perfectly conducting walls placed at $x = 0$ and $x = a$, where the electric field must vanish. In fact, this standing behavior is formally identical to the standing waves observed in a vibrating string with fixed ends: $y(0,t) = 0$ and $y(L,t) = 0$. Similarly, the existence of nodal planes (in which the electric field is zero) is inferred from the term $\sin(k_c x)$. Their position can be computed setting $\sin(k_c x) = 0$, which is equivalent to $k_c x = n\pi$, where n is an integer number. In addition, it is well known that a string with fixed ends cannot vibrate with arbitrary frequencies. We may therefore expect that the frequencies associated to the cutoff wave number are also quantized.

3.2.3. Dispersion Relation and Cutoff Frequencies

From the definition of k_g and k_c, it can be easily proven (applying the fundamental Pythagorean trigonometric identity) that:

$$k_g^2 + k_c^2 = k^2 \tag{3.15}$$

At this point, it is convenient to define the cutoff frequency as $\omega_c \equiv ck_c$ and recall the relation between angular frequency and wavenumber in free space, $\omega = ck$. Inserting this expression into Equation 3.15 yields:

$$\omega = \sqrt{\omega_c^2 + c^2 k_g^2} \tag{3.16}$$

This expression relating ω and k_g, whose precise form depends on the type of guiding structure and the particular propagating mode, is known as dispersion relation. In general, dispersion relations for confined waves differ from the relationship between ω and k in free space, $\omega = ck$, which states that these two quantities are proportional. As a general rule, this proportionality is not satisfied for guided waves. For high frequencies (and k_g-values), the square of the cutoff frequency is negligible (compared to the other squared term) and the frequency is approximately proportional to the propagation wave number along the guide. This can also be confirmed drawing the function $\omega(k_g)$ given by Equation 3.16, which asymptotically tends to:

$$\omega = ck_g \tag{3.17}$$

Now, let us focus on the cutoff frequency. As mentioned previously, the tangential component of the electric field must be zero at the sides of the perfectly conducting walls, which leads to the boundary conditions given by Equations 3.6. The reader should note that, so far, we have only applied the condition for $x = 0$, but the condition $E_z(a, y, t) = 0$ has not been imposed yet. Such a condition will be satisfied for any y and t if:

$$2E_0 \sin(k_c a)\sin(\omega t - k_g y + \phi) = 0 \tag{3.18}$$

This will be true if $\sin(k_c a) = 0$, which implies:

$$k_c a = n\pi \qquad n = 1,2,3,... \tag{3.19}$$

This is equivalent to:

$$\omega_c = n\pi \frac{c}{a} \qquad n = 1,2,3,... \tag{3.20}$$

And recalling the relationship between frequency and angular frequency, we have:

$$f_c = \frac{\omega_c}{2\pi} = n\frac{c}{2a} \qquad n = 1,2,3,... \tag{3.21}$$

This equation is formally identical to the equation that frequencies of a vibrating string with fixed ends must obey:

$$f = n\frac{v}{2L} \qquad n = 1,2,3,... \tag{3.22}$$

In any case, Equations 3.19, 3.20 and 3.21 tell us that cutoff wave numbers, cutoff angular frequencies and cutoff frequencies are quantized. For any n-value, we have a cutoff frequency as well as a propagation mode, which will be denoted as TE_n. You should note that TE_0 is a trivial mode whose electric field is zero everywhere (check it setting $k_c = 0$ in Equation 3.14), which cannot be considered a propagation mode. For that reason, $n = 0$ was deliberately omitted in Equations 3.19-3.22.

From Equation 3.16, we additionally conclude that $\omega \geq \omega_c$. In other words, the frequency has a lower bound that depends on the propagation mode. A wave will propagate along the guide if its frequency is greater than the cutoff frequency. Otherwise, the wave will attenuate exponentially along the guide direction.

3.2.4. The Magnetic Field and the Poynting Vector

At this point, it is worth computing the magnetic field and the time-averaged Poynting vector. This can be done in an easy and rapid way by applying Faraday's law (although this is not the only feasible route):

$$\vec{\nabla} \times \vec{E} = -\frac{\partial \vec{B}}{\partial t} \tag{3.23}$$

Recalling that the curl can be expanded with the help of a determinant and keeping in mind that the electric field only has z-component yields:

$$\vec{\nabla} \times \vec{E} = \begin{vmatrix} \hat{i} & \hat{j} & \hat{k} \\ \frac{\partial}{\partial x} & \frac{\partial}{\partial y} & 0 \\ 0 & 0 & E_z \end{vmatrix} = \hat{i}\frac{\partial E_z}{\partial y} - \hat{j}\frac{\partial E_z}{\partial x} = -\frac{\partial \vec{B}}{\partial t} \tag{3.24}$$

Solving for the magnetic field gives:

$$\vec{B} = \int \left(-\hat{i}\frac{\partial E_z}{\partial y} + \hat{j}\frac{\partial E_z}{\partial x} \right) dt \tag{3.25}$$

Dealing with harmonic functions, partial derivatives and integrals are really easy to perform. The reader can check that:

$$\vec{B}(x,y,t) = 2E_0(k_g/\omega)\sin(k_c x)\sin(\omega t - k_g y + \phi)\hat{i}$$
$$-2E_0(k_c/\omega)\cos(k_c x)\cos(\omega t - k_g y + \phi)\hat{j}$$

(3.26)

Note that the magnetic field does have component in the guiding direction (the y-component). It is quite instructive to visualize the direction of this field drawing the corresponding field lines, which are plotted in Figure 3.3 for a TE_1 mode.

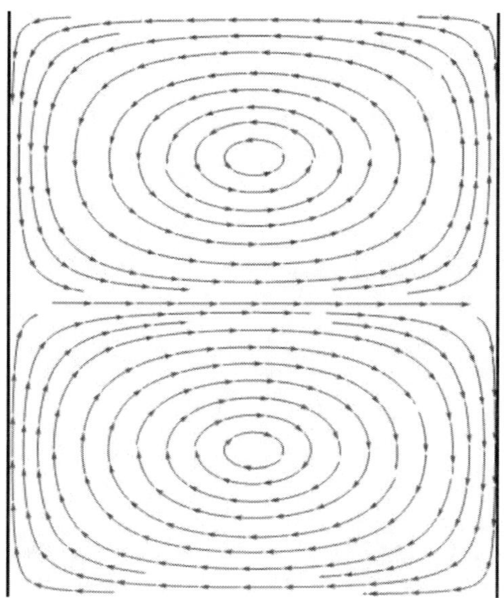

Figure 3.3. Magnetic field lines for TE_1 modes. At any given point these lines are tangential to the magnetic field. Figure generated by Mathematica 8, a registered trademark of Wolfram Research, Inc.

The length of the portion of the guide drawn in the propagation direction equals $\lambda_g = 2\pi/k_g$. In other words, a pattern of only one cycle has been plotted. This pattern is periodically replicated along the guiding direction. We should also keep in mind that the entire picture shows an instantaneous configuration that moves along the y-direction. As can be seen the pattern consists in two groups of closed loops circulating in opposite directions. Recall that magnetic field lines do not originate and terminate on any point. It should be also stressed that, near the guide walls, these lines are tangential to the perfectly conducting surface. This property will be justified on physical grounds in section 3.5.1.

From the expressions for the electric and magnetic fields (Equation 3.14 and 3.26), the Poynting vector turns out to be:

$$\vec{S} = \frac{\vec{E} \times \vec{B}}{\mu_0} = \frac{1}{\mu_0} \begin{vmatrix} \hat{i} & \hat{j} & \hat{k} \\ 0 & 0 & E_z \\ B_x & B_y & 0 \end{vmatrix} = \frac{-\hat{i}E_z B_y + \hat{j}E_z B_x}{\mu_0} \tag{3.27}$$

As can be seen, the Poynting vector depends on x, y and t and has x- and y- components. However, it is more interesting to compute the time-averaged Poynting vector, which gives us the average flux of power per unit area (intensity), as mentioned in the previous chapter. Time averaging is a linear operation. Thus:

$$\langle \vec{S} \rangle = \frac{-\hat{i}\langle E_z B_y \rangle + \hat{j}\langle E_z B_x \rangle}{\mu_0} \tag{3.28}$$

The first average is computed as follows:

$$\langle E_z B_y \rangle =$$
$$= \langle 2E_0 \sin(\omega t - k_g y + \phi) \cdot (-2E_0(k_c/\omega)\cos(k_c x)\cos(\omega t - k_g y + \phi)) \rangle = \tag{3.29}$$
$$- E_0^2(k_c/\omega) \cdot 2\sin(k_c x)\cos(k_c x)\langle 2\sin(\omega t - k_g y + \phi)\cos(\omega t - k_g y + \phi) \rangle$$

Taking into account that $\sin(2\alpha) = 2\sin\alpha\cos\alpha$ yields:

$$\langle E_z B_y \rangle = -E_0^2(k_c/\omega) \cdot \sin(2k_c x)\langle \sin(2\omega t - 2k_g y + 2\phi) \rangle \tag{3.30}$$

The time average of a harmonic function is zero because it regularly oscillates between -1 and $+1$. Thus:

$$\left\langle E_z B_y \right\rangle = 0 \tag{3.31}$$

We therefore end up with:

$$\left\langle \vec{S} \right\rangle = \frac{\hat{j}\left\langle E_z B_x \right\rangle}{\mu_0} \tag{3.32}$$

Bringing to mind that the time average of the square of a sine function is ½, it can be easily shown that:

$$\left\langle \vec{S} \right\rangle = \frac{4E_0^2 \sin^2(k_c x)k_g \left\langle \sin^2(\omega t - k_g y + \phi) \right\rangle}{\mu_0 \omega} \hat{j} =$$
$$= \frac{2E_0^2 \sin^2(k_c x)k_g}{\mu_0 \omega} \hat{j} \tag{3.33}$$

Equivalently, the intensity is given by:

$$I = \frac{2E_0^2 \sin^2(k_c x)k_g}{\mu_0 \omega} = 2E_0^2 \sin^2(k_c x)\varepsilon_0 c \cos\theta \tag{3.34}$$

On average, the energy only travels in the y-direction, which is the guiding direction in this case. This agrees with the fact that the wave is stationary in the transverse direction.

3.2.5. Phase Velocity and Group Velocity

It is well known that the velocity of plane harmonic waves propagating through free space is ω/k. This quotient can also be interpreted geometrically as follows. Let us imagine the electromagnetic wave $E_z = E_0 \sin(\omega t - ky)$ in open space and consider the plane (or wavefront) whose phase is zero and therefore satisfies $\omega t - ky = 0$. at $t = 0$ this zero-phase plane is located at $y = 0$. After certain time t, the zero-phase wavefront is located at $y = \omega t/k$, so this plane of constant phase (or any other) moves with the speed $c = \omega/k$. For that reason, this quotient is known as *phase* velocity (v_p). For free electromagnetic waves the phase velocity is identical to the speed of light.

What is the phase velocity for a guided wave? In this case the wave equation is:

$$E_z = 2E_0 \sin(k_c x)\sin(\omega t - k_g y + \phi) \tag{3.35}$$

Thus the phase velocity is:

$$v_p = \omega/k_g = \omega/k\cos\theta = c/\cos\theta \tag{3.36}$$

Given that the cosine of any angle is not greater than 1, Equation 3.36 suggests that the phase velocity of a guided wave is greater than the speed of light. This conclusion might seem striking for some readers but it does not violate the theory of relativity because the energy and the signal do not travel with this speed. The phase velocity is a formal concept rather than a measurable physical quantity.

First, let us show that the velocity of the signal is not greater than the speed of light using again the model of successive reflections. This intuitive picture suggests that, bouncing left and right with the speed of light, the signal cannot propagate more rapidly than light in the guiding direction. Consider one of the bounces in the path going from point A to point B (see Figure 3.4).

If Δs is the distance between A and B, it is quite easy to show that the distance traveled by the rays bouncing left and right (Δs_{real}) is greater:

$$\Delta s_{real} = \frac{\Delta s/2}{\cos\theta} + \frac{\Delta s/2}{\cos\theta} = \frac{\Delta s}{\cos\theta} \tag{3.37}$$

It takes the time:

$$\Delta t = \frac{\Delta s_{real}}{c} = \frac{\Delta s}{c\cos\theta} \tag{3.38}$$

The velocity of propagation of the signal along the guide, usually known as *group* velocity, is therefore given by:

$$v_g = \frac{\Delta s}{\Delta t} = \frac{\Delta s}{\left(\dfrac{\Delta s}{c\cos\theta}\right)} = c\cos\theta \tag{3.39}$$

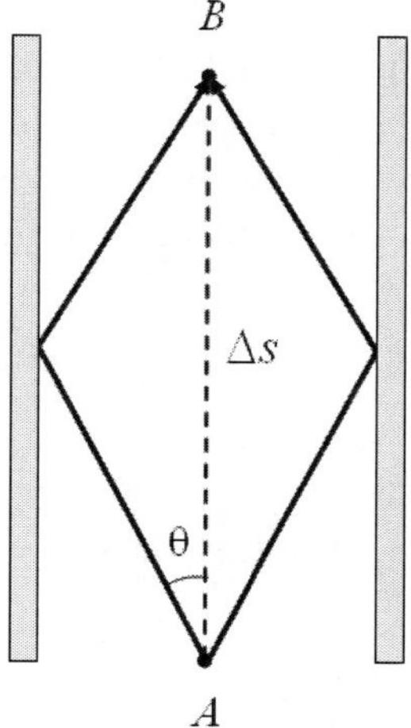

Figure 3.4. Bouncing left and right with the speed of light, the signal cannot propagate more rapidly than light in the guiding direction.

As can be concluded, this velocity is not greater than the speed of light.

Now, let us show that the energy propagates with the same velocity. We have previously proved that the intensity of a plane wave traveling through free space is

$$I = \varepsilon_0 c\left\langle E_z^2 \right\rangle = \tfrac{1}{2}\varepsilon_0 c E_0^2 \tag{3.40}$$

In this section we have also calculated the intensity of a guided wave:

$$I = \varepsilon_0 c\cos\theta \cdot 2E_0^2 \sin^2(k_c x) \tag{3.41}$$

The term $2E_0^2 \sin^2(k_c x)$ appearing in this equation can be easily identified as the time average of the square of the amplitude of the electric field, $\left\langle E_z^2 \right\rangle$, which leads to:

$$I = \varepsilon_0 c \cos\theta \left\langle E_z^2 \right\rangle \qquad (3.42)$$

Comparing this result (for guided waves) and Equation 3.40 (for free waves), we can conclude that the apparent velocity of the energy flux for a guided wave is $c\cos\theta$, which is the group velocity.

To end this section we should highlight two results derived for the parallel-plate guide but with general validity. First, note that:

$$v_p v_g = c^2 \qquad (3.43)$$

Second, consider the dispersion relation previously found $\omega = \sqrt{\omega_c^2 + c^2 k_g^2}$. It can easily be shown from it that:

$$v_g = \frac{d\omega}{dk_g} \qquad (3.44)$$

We again insist on the general character of Equations 3.43 and 3.44.

3.3. THE PARALLEL-PLATE WAVEGUIDE: TM AND TEM MODES

3.3.1. A Boundary Condition for the Magnetic Field

In the previous section we have studied the propagation of an electromagnetic wave in a parallel-plate waveguide assuming that the electric field had no longitudinal component in the guiding direction (y-direction in our example).

Now, let us consider the case in which the field lacking longitudinal component is the magnetic field (TM modes). This new situation can be analyzed to a great extent by following the same line of reasoning used above. Here, we will pay attention to those aspects that differ in both treatments.

Again there are two plane waves obliquely bouncing left and right. The magnetic field must be perpendicular to the directions of these waves and to the y-direction. This is possible if this field only has z-component (see Figure 3.5).

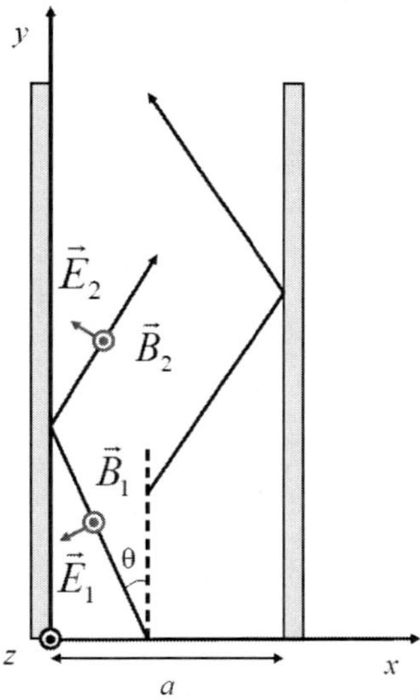

Figure 3.5. Successive reflections in a parallel-plate waveguide. The wave resulting from the superposition of wave 1 and wave 2 propagates along the +y-direction. Note that the magnetic field has no component in the direction of propagation (TM mode).

First, we will compute the magnetic field of the wave resulting from the superposition of plane waves traveling in two directions. Then, the electric field will be calculated. The magnetic fields associated to the waves coming from the right and the left walls will be identified with subscripts 1 and 2, respectively. Their expressions are formally identical to Equation 3.5 for the electric field:

$$B_{1z} = B_{10}\sin(\omega t + kx\sin\theta - ky\cos\theta + \phi_1)$$
$$B_{2z} = B_{20}\sin(\omega t - kx\sin\theta - ky\cos\theta + \phi_2)$$

$$(3.45)$$

To add these fields, we should previously know the relationships linking both amplitudes and both phase constants. As you can guess, boundary conditions play an important role in this task and the rest of the discussion. The tangential component of the electric field must again be zero at the surface of the perfectly conducting walls.

According to Figure 3.5, the electric field has now two components, but only the y-component is tangential to the surface of the walls. Thus

$$E_y(0, y, t) = 0$$
$$E_y(a, y, t) = 0$$

$$(3.46)$$

In this case, however, these boundary conditions are not very useful because we are dealing with the magnetic field. Thus boundary conditions for it must be derived. This task will be carried out applying the differential form of Ampère-Maxwell's law in the immediate neighborhood of the walls:

$$\vec{\nabla} \times \vec{B} \equiv \begin{vmatrix} \hat{i} & \hat{j} & \hat{k} \\ \frac{\partial}{\partial x} & \frac{\partial}{\partial y} & 0 \\ 0 & 0 & B_z \end{vmatrix} = \frac{\partial B_z}{\partial y} \hat{i} - \frac{\partial B_z}{\partial x} \hat{j} =$$

$$= \mu_0 \varepsilon_0 \left(\frac{\partial E_x}{\partial t} \hat{i} + \frac{\partial E_y}{\partial t} \hat{j} \right)$$

$$(3.47)$$

Focusing on the y-component we have:

$$-\frac{\partial B_z}{\partial x} = \mu_0 \varepsilon_0 \frac{\partial E_y}{\partial t}$$

$$(3.48)$$

Inserting Equations 3.46 into Equation 3.48 yields the boundary conditions we are looking for:

$$\left. \frac{\partial B_z}{\partial x} \right|_{x=0} = 0$$

$$\left. \frac{\partial B_z}{\partial x} \right|_{x=a} = 0$$

$$(3.49)$$

3.3.2. Fields and Cutoff Frequencies in TM Modes

Now, the first of these boundary conditions will be applied to $B_z = B_{1z} + B_{2z}$, which leads to:

$$B_{10} k \sin\theta \cos(\omega t - ky\cos\theta + \phi_1)$$
$$- B_{20} k \sin\theta \cos(\omega t - ky\cos\theta + \phi_2) = 0 \tag{3.50}$$

This expression can be simplified by factoring:

$$B_{10}\cos(\omega t - ky\cos\theta + \phi_1) - B_{20}\cos(\omega t - ky\cos\theta + \phi_2) = 0 \tag{3.51}$$

This subtraction will be identically zero for any y and t if:

$$B_{10} = B_{20} \equiv B_0$$
$$\phi_2 = \phi_1 \equiv \phi \tag{3.52}$$

Note that both amplitudes must be the same. This also happened for the electric field in TE modes. However, the phase constants must also be identical (whereas they differed in π radians for TE modes). Within these conditions, $B_z = B_{1z} + B_{2z}$ can be computed. Applying a similar way of proceeding, it can be shown that:

$$B_z(x, y, t) = 2B_0 \cos(k_c x)\sin(\omega t - k_g y + \phi) \tag{3.53}$$

This expression resembles the result that we obtained for the electric field in TE modes, but the function $\sin(k_c x)$ has now been replaced by $\cos(k_c x)$. The definitions for the propagation wave number along the guide direction and the cutoff wave number are the same. Thus, the dispersion relation remains unaltered. The cutoff frequencies are calculated from the boundary condition at $x = a$ (see Equations 3.49). After taking the partial derivative of B_z with respect to x, evaluating at $x = a$ and equaling zero, we obtain:

$$k_c a = n\pi \qquad n = 0,1,2,3,... \tag{3.54}$$

And also:

$$\omega_c = n\pi \frac{c}{a} \qquad n = 0,1,2,3,...$$

(3.55)

The mode of propagation associated to a given n-value is denoted as TM_n. Finally, the electric field can be computed from Equation 3.47 (Ampère-Maxwell's law):

$$\vec{E}(x,y,t) = -2B_0 c^2 (k_g / \omega)\cos(k_c x)\sin(\omega t - k_g y + \phi)\hat{i}$$
$$- 2B_0 c^2 (k_c / \omega)\sin(k_c x)\cos(\omega t - k_g y + \phi)\hat{j}$$

(3.56)

Figure 3.6. Electric field lines for TM_1 modes. Figure generated by Mathematica 8, a registered trademark of Wolfram Research, Inc.

Figure 3.6 shows the field lines corresponding to this electric field for a TM_1 mode. Again, we have plotted only one cycle in the guiding direction (the length of the portion of the guide drawn in the y-direction is $\lambda_g = 2\pi / k_g$). This instantaneous pattern is periodically replicated (and moves) along the guiding direction.

As can be seen, the field lines begin and end on the guide plates. Recall that electric field lines emanate from positive charges and terminate on negative charges. This means that the plate surfaces must be locally charged (but not globally). It should be also stressed that, near the guide walls, these lines are perpendicular to the perfectly conducting surface, which agrees with the boundary condition deduced in section 3.2.1.

3.3.3. TEM Modes

To begin with, it must be stressed a small but important difference between TE and TM modes: $n = 0$ is feasible for TM modes due to the presence of $\cos(k_c x)$ in Equation 3.53, which does not vanish for $n = 0$. This implies that $k_c = 0$, $k_g = k$, and $\theta = 0$. Inserting these values into Equations 3.53 and 3.56 gives:

$$\vec{B}(x, y, t) = 2B_0 \sin(\omega t - k_g y + \phi)\hat{k} \tag{3.57}$$

$$\vec{E}(x, y, t) = -2B_0 c \sin(\omega t - k_g y + \phi)\hat{i} \tag{3.58}$$

In addition, we have $\omega = ck_g$ because $\omega_c = 0$. We therefore conclude that:

- Both fields are perpendicular to the direction of propagation. For this reason, the TM_0 mode is commonly known as TEM mode.
- The frequency and the wave number along the guide direction are identical to the values for free space.
- The frequency and the wave number along the guide direction are proportional and the proportionality constant is the speed of light in empty space.
- As $\theta = 0$, both group velocity and phase velocity are identical to the speed of light in empty space.

In summary: guided electromagnetic waves propagate in TEM modes as if they were not confined (or, equivalently, in free space). This should not be surprising because for $\theta = 0$ waves do not bounce between the walls.

Finally, it should be stressed that TEM modes admit a physical description in terms of current and voltage whereas TE_n and TM_n modes (with $n > 0$) can only be described in terms of fields (as done in this book). The proof of this outstanding property of TEM modes goes beyond the scope of this book but can be found in advanced electromagnetism treatments. For TEM modes, the term *transmission line* is preferred rather than waveguide.

3.4. Rectangular Waveguide: TE Modes

3.4.1. Electric Field

In this section we discuss in detail the case of a rectangular hollow waveguide with perfectly conducting walls, whose cross section is shown in Figure 3.7. Without loss of generality, we may assume that the lengths a, b of the inner sides satisfy $b \leq a$. The fields are now confined in x- and y-directions and propagate along the +z-direction. This implies that $E_z = 0$ for transverse electric (TE) modes. However, the other components will have a non-trivial dependence on the transverse coordinates x and y, which is calculated as follows.

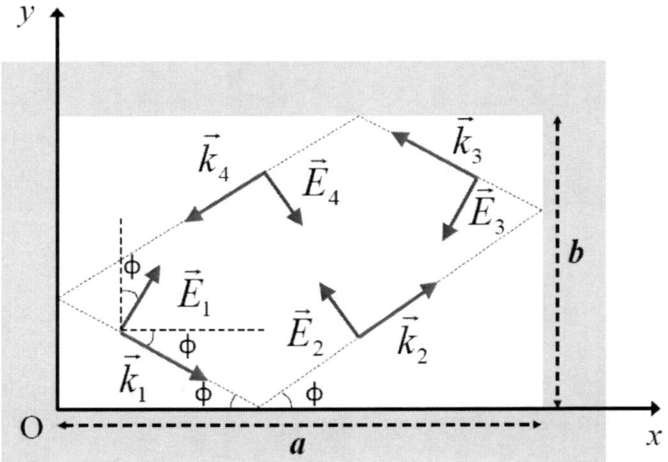

Figure 3.7. Projection of the successive reflections in a rectangular waveguide onto the xy-plane. The wave resulting from the superposition of different waves propagates along the +z-direction (not shown). The electric field has no component in the direction of propagation (TE mode). However, it should be stressed that the arrows representing wave vectors are only projections onto the xy-plane and have z-component.

In this case, we have the superposition of four waves with different wave vectors (listed below), as a result of the successive reflections at the perfect conducting walls. Figure 3.7 shows a projection of this four wave vectors onto the *xy* plane. Please, keep in mind that in this section the wave propagates along the *z*-direction, which is perpendicular to the sheet of paper.

If k_x, k_y and k_z are the *absolute* values of the Cartesian components of one of these wave vectors, we can write:

$$\vec{k}_1 = k_x\hat{i} - k_y\hat{j} + k_z\hat{k}$$
$$\vec{k}_2 = k_x\hat{i} + k_y\hat{j} + k_z\hat{k}$$
$$\vec{k}_3 = -k_x\hat{i} + k_y\hat{j} + k_z\hat{k}$$
$$\vec{k}_4 = -k_x\hat{i} - k_y\hat{j} + k_z\hat{k}$$

$$(3.59)$$

Figure 3.7 also shows the electric field corresponding to each wave. From this figure, we can easily conclude that:

$$E_{1x} = E_1(\vec{r},t)\sin\phi = E_0 \sin(\vec{k}_1 \cdot \vec{r} - \omega t)\sin\phi$$
$$E_{2x} = -E_2(\vec{r},t)\sin\phi = -E_0 \sin(\vec{k}_2 \cdot \vec{r} - \omega t)\sin\phi$$
$$E_{3x} = -E_3(\vec{r},t)\sin\phi = -E_0 \sin(\vec{k}_3 \cdot \vec{r} - \omega t)\sin\phi$$
$$E_{4x} = E_4(\vec{r},t)\sin\phi = E_0 \sin(\vec{k}_4 \cdot \vec{r} - \omega t)\sin\phi$$

$$(3.60)$$

In relation to Figure 3.7, it should be mentioned that the relative orientations of the electric field before and after each reflection have not been randomly chosen. These fields have been drawn to explicitly illustrate that the tangential component of the electric field resulting from the incident and reflected waves vanishes at a perfectly conducting wall. For instance, E_{1x} and E_{2x} must have opposite signs to satisfy that the *x*-component of the electric field resulting from waves 1 and 2 is zero at the wall located at $y = 0$. Analogously, E_{4y} and E_{1y} must also have opposite signs to satisfy that the *y*-component of the electric field is zero when the electric field is reflected at the wall located at $x = 0$. This change of sign is similar to the addition of π radians in Equation 3.9, when the TE mode in a parallel-plate waveguide was analyzed.

You should also bear in mind that we have drawn the electric field vectors corresponding to four points only, but such fields fill the whole space inside the guide. Thus at any point the x-component of the electric field is given by the sum of the four x-components written above:

$$E_x(\vec{r},t) = E_{1x} + E_{2x} + E_{3x} + E_{4x} =$$
$$E_0 \sin\phi[\sin(k_x x - k_y y + k_z z - \omega t) - \sin(k_x x + k_y y + k_z z - \omega t) \quad (3.61)$$
$$- \sin(-k_x x + k_y y + k_z z - \omega t) + \sin(-k_x x - k_y y + k_z z - \omega t)]$$

In this expression there are two subtractions of sine functions, which can be performed applying the identity:

$$\sin\alpha - \sin\beta = 2\sin\left(\frac{\alpha - \beta}{2}\right)\cos\left(\frac{\alpha + \beta}{2}\right) \qquad (3.62)$$

Then we can conclude that:

$$E_x(\vec{r},t) = 2E_0 \sin\phi[\sin(-k_y y)\cos(k_x x + k_z z - \omega t)$$
$$+ \sin(-k_y y)\cos(-k_x x + k_z z - \omega t)] = \qquad (3.63)$$
$$= -2E_0 \sin\phi\sin(k_y y)[\cos(k_x x + k_z z - \omega t) + \cos(-k_x x + k_z z - \omega t)]$$

Now, the addition of two cosines can be carried out applying the identity:

$$\cos\alpha + \cos\beta = 2\cos\left(\frac{\alpha - \beta}{2}\right)\cos\left(\frac{\alpha + \beta}{2}\right) \qquad (3.64)$$

In this way, it can be proved that:

$$E_x(\vec{r},t) = -4E_0 \sin\phi\sin(k_y y)\cos(k_x x)\cos(k_z z - \omega t) \qquad (3.65)$$

Recalling that Figure 3.7 shows the projection of the wave vectors onto the xy-plane, it can be easily shown that:

$$\sin\phi = \frac{k_y}{\sqrt{k_x^2 + k_y^2}} \tag{3.66}$$

Thus the x-component of the electric field can be written as:

$$E_x(\vec{r},t) = -4E_0 \frac{k_y}{\sqrt{k_x^2 + k_y^2}} \cos(k_x x)\sin(k_y y)\cos(k_z z - \omega t) \tag{3.67}$$

The y-component of the electric field can be calculated from the y-components of the four waves:

$$\begin{aligned}
E_{1y} &= E_1(\vec{r},t)\cos\phi = E_0 \sin(\vec{k}_1 \cdot \vec{r} - \omega t)\cos\phi \\
E_{2y} &= E_2(\vec{r},t)\cos\phi = E_0 \sin(\vec{k}_2 \cdot \vec{r} - \omega t)\cos\phi \\
E_{3y} &= -E_3(\vec{r},t)\cos\phi = -E_0 \sin(\vec{k}_3 \cdot \vec{r} - \omega t)\cos\phi \\
E_{4y} &= -E_4(\vec{r},t)\cos\phi = -E_0 \sin(\vec{k}_4 \cdot \vec{r} - \omega t)\cos\phi
\end{aligned} \tag{3.68}$$

Following a similar reasoning, you can prove that (see problem section):

$$E_y(\vec{r},t) = 4E_0 \frac{k_x}{\sqrt{k_x^2 + k_y^2}} \sin(k_x x)\cos(k_y y)\cos(k_z z - \omega t) \tag{3.69}$$

You can also show that this electric field satisfies the differential form of Gauss' law. In any case, all these expressions confirm the existence of a traveling wave in the z-direction and standing waves in the transverse directions, as expected. Thus k_z can be identified with the wave number in the guiding direction, k_g in our notation, and β for other authors.

3.4.2. Cutoff Frequencies and Dispersion Relation

You can easily check that this electric field also satisfied the boundary conditions at the perfectly conducting walls located at $x = 0$ and $y = 0$, i.e., the components parallel to such walls must be zero:

$$E_y(0, y, z, t) = 0$$
$$E_x(x, 0, z, t) = 0 \tag{3.70}$$

Obviously, similar boundary conditions must be satisfied at $x = a$ and $y = b$:

$$E_y(a, y, z, t) = 0$$
$$E_x(x, b, z, t) = 0 \tag{3.71}$$

According to Equations 3.67 and 3.69, this will happen if:

$$k_x = n\pi / a \qquad n = 0, 1, 2, 3, \ldots$$
$$k_y = m\pi / b \qquad m = 0, 1, 2, 3, \ldots \tag{3.72}$$

In other words, the values of the transverse wave numbers k_x and k_y are quantized and we refer to the corresponding modes as TE$_{nm}$. As $k^2 = k_x^2 + k_y^2 + k_z^2 = k_x^2 + k_y^2 + k_g^2$ and k is fixed (wave number in free space), k_g-values are also quantized:

$$k_g = \sqrt{k^2 - k_x^2 - k_y^2} = \sqrt{k^2 - \left(\frac{n\pi}{a}\right)^2 - \left(\frac{m\pi}{b}\right)^2} \tag{3.73}$$

Defining $k_c \equiv \sqrt{(n\pi/a)^2 + (n\pi/b)^2}$, $\omega_c \equiv ck_c$ and recalling $\omega = ck$, this expression leads to a dispersion relation identical to that obtained for the parallel-plate guide:

$$\omega = \sqrt{\omega_c^2 + c^2 k_g^2} \tag{3.74}$$

The cutoff frequencies can again be defined as:

$$f_c = \frac{\omega_c}{2\pi} = c\sqrt{\left(\frac{n}{2a}\right)^2 + \left(\frac{m}{2b}\right)^2} \qquad n, m = 0, 1, 2, 3, \ldots \tag{3.75}$$

Propagation only takes place above the cutoff frequency. Below this threshold, the wave will attenuate exponentially along the guiding direction. In practice, it is desirable to operate in a frequency range that ensures that only one mode can propagate. If several modes propagate simultaneously, one can hardly control which modes are actually carrying the transmitted signal. If we arrange the cutoff frequencies in increasing order, $f_{c1} < f_{c2} < f_{c3} < ..$, the single-mode operation can be ensured if the frequency is restricted to $f_{c1} < f < f_{c2}$. In such a way, only the lowest mode will propagate. This interval is usually known as *operating bandwidth* of the guide.

Finally you should note that mode TE_{00} is the trivial solution $\vec{E}(x, y, z, t) = 0$, which means that nothing is propagating along the guide.

3.4.3. Magnetic Field

The magnetic field can be calculated from the electric field with the help of the differential form of Faraday's law, as done with the parallel-plate waveguide. You will obtain the following Cartesian components:

$$B_x(\vec{r}, t) = -4E_0 \frac{k_x k_z}{\omega\sqrt{k_x^2 + k_y^2}} \sin(k_x x)\cos(k_y y)\cos(k_z z - \omega t)$$

$$B_y(\vec{r}, t) = -4E_0 \frac{k_y k_z}{\omega\sqrt{k_x^2 + k_y^2}} \cos(k_x x)\sin(k_y y)\cos(k_z z - \omega t) \quad (3.76)$$

$$B_z(\vec{r}, t) = 4E_0 \frac{\sqrt{k_x^2 + k_y^2}}{\omega} \cos(k_x x)\cos(k_y y)\sin(k_z z - \omega t)$$

This magnetic field also satisfies the differential form of Gauss's law for magnetism (check it yourself). In many textbooks of advanced electromagnetism, it is usual to write the Cartesian components of electric and magnetic fields of TE_{nm} modes as a function of the amplitude of B_z (the only longitudinal component). Defining $B_{z0} \equiv 4E_0\sqrt{k_x^2 + k_y^2}\,/\,\omega$ gives:

$$E_x(\vec{r},t) = -B_{z0}\,\frac{\omega k_y}{k_x^2 + k_y^2}\cos(k_x x)\sin(k_y y)\cos(k_g z - \omega t)$$

$$E_y(\vec{r},t) = B_{z0}\,\frac{\omega k_x}{k_x^2 + k_y^2}\sin(k_x x)\cos(k_y y)\cos(k_g z - \omega t)$$

$$B_x(\vec{r},t) = -B_{z0}\,\frac{k_x k_g}{k_x^2 + k_y^2}\sin(k_x x)\cos(k_y y)\cos(k_g z - \omega t) \qquad (3.77)$$

$$B_y(\vec{r},t) = -B_{z0}\,\frac{k_y k_g}{k_x^2 + k_y^2}\cos(k_x x)\sin(k_y y)\cos(k_g z - \omega t)$$

$$B_z(\vec{r},t) = B_{z0}\cos(k_x x)\cos(k_y y)\sin(k_g z - \omega t)$$

Working with this expression, do not forget that k_x and k_y are quantized, and their values are given by Equation 3.72. It would be instructive to approximately draw some field lines corresponding to Equations 3.77. It should be mentioned, however, that the magnetic field has components in the three directions of space, which makes this task quite difficult. Fortunately, the electric field has only two components. Thus its field lines can be properly plotted in 2-D. Figure 3.8 shows these field lines for TE_{11} and TE_{21} modes (at a given time).

These patterns satisfy the rules of thumb for electric field lines discussed previously and corroborated for the parallel-plate guide:

i)   The field lines begin and end on the guide walls. Recall that electric field lines emanate from positive charges and terminate on negative charges. Thus the guide surface must be locally charged (but not globally).

ii) Near the guide walls, the lines are perpendicular to the perfectly conducting surface, in agreement with the boundary condition deduced in section 3.2.1.

In addition, note how the spatial pattern changes when m varies from 1 to 2.

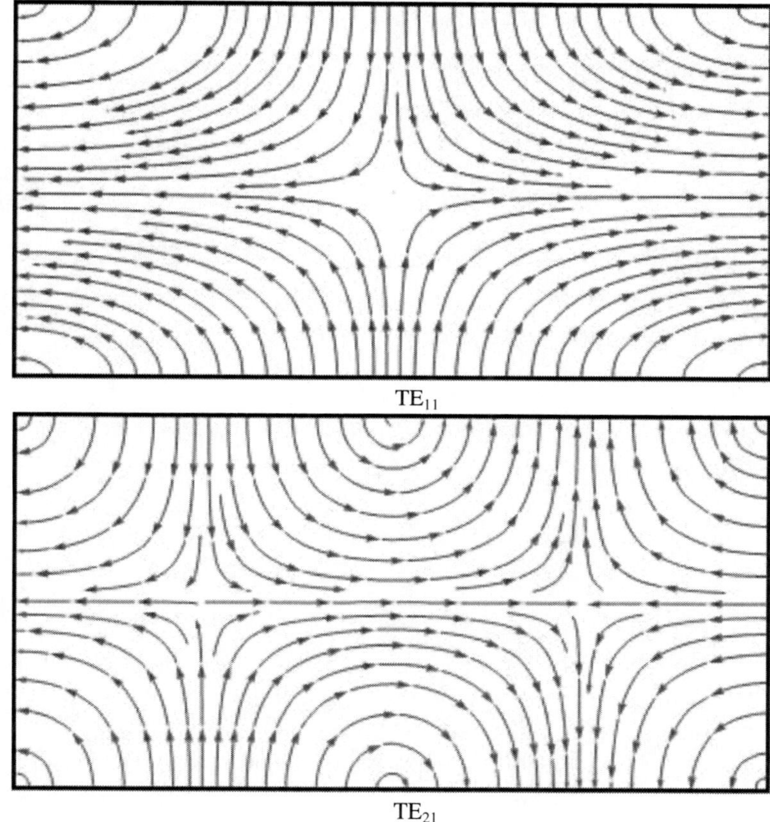

Figure 3.8. Electric field lines for TE$_{11}$ and TE$_{21}$ modes for a rectangular waveguide with $a=2b$. Figure generated by Mathematica 8, a registered trademark of Wolfram Research, Inc.

3.4.4. Transmitted Power

The expressions of the fields can be useful for different tasks. For instance, we can use them to compute the time-averaged power transmitted in the TE$_{10}$ mode, for which $k_x = \pi/a$ and $k_y = 0$. Inserting these values into Equations 3.77 and setting $E_{0y} \equiv B_0\omega/k_x$ gives:

$$E_x = 0$$

$$E_y = E_{0y}\sin(\pi x/a)\cos(k_g z - \omega t)$$

$$E_z = 0$$

$$B_x = -E_{0y}\frac{k_g}{\omega}\sin(\pi x/a)\cos(k_g z - \omega t) \tag{3.78}$$

$$B_y = 0$$

$$B_z = E_{0y}\frac{\pi/a}{\omega}\cos(\pi x/a)\sin(k_g z - \omega t)$$

Note that E_{0y} is the amplitude of the resulting electric field. From these components, the Poynting vector is calculated as:

$$\vec{S} = \frac{\vec{E}\times\vec{B}}{\mu_0} = \frac{1}{\mu_0}\begin{vmatrix}\hat{i} & \hat{j} & \hat{k}\\ 0 & E_y & 0\\ B_x & 0 & B_z\end{vmatrix} = \frac{\hat{i}E_y B_z - \hat{k}E_y B_x}{\mu_0} \tag{3.79}$$

But the time-averaged Poynting vector only has component in the guiding direction, as we have explicitly shown for the parallel-plate guide. Thus:

$$\langle\vec{S}\rangle = \frac{-\hat{k}\langle E_y B_x\rangle}{\mu_0} =$$

$$= \frac{-\hat{k}\langle E_{0y}^2(k_g/\omega)\sin^2(\pi x/a)\cos^2(k_g z - \omega t)\rangle}{\mu_0} \tag{3.80}$$

Recalling that the time average of the square of a harmonic function is 1/2, the intensity is:

$$I = \langle|\vec{S}|\rangle = \frac{E_{0y}^2 k_g \sin^2(\pi x/a)}{2\omega\mu_0} \tag{3.81}$$

The transmitted power is obtained by integrating this intensity over the cross-sectional area of the guide:

$$P = \int_0^b \int_0^a I\,dxdy = \frac{E_{0y}^2 k_g}{2\mu_0 \omega} \int_0^b \int_0^a \sin^2(\pi x/a)\,dxdy =$$

$$= \frac{E_{0y}^2 k_g}{2\mu_0 \omega} \frac{ba}{2} = \frac{E_{0y}^2 k_g ba}{4\mu_0 \omega}$$

(3.82)

3.5. Rectangular Waveguide. TM Modes

3.5.1. Another Boundary Condition for the Magnetic Field

The calculation of the electromagnetic field based on successive reflections at the perfectly conducting walls can also be applied to transverse magnetic modes ($B_z = 0$).

Before facing this task, it would be helpful to derive a new boundary condition for the magnetic field from Gauss' law. First, consider a small 'pillbox' enclosing a tiny piece of the boundary between free space and a perfectly conducting wall (see Figure 3.9).

The flat sides of the box are parallel to the wall. The flux of the magnetic field through the surface of this pillbox can be computed breaking the integral over the closed surface into three parts: Over the flat sides S_1 and S_2 and over the curved surface (S_3).

The magnetic field inside perfectly conducting media is zero. Thus the integral over S_2 is also zero. In the limit that the thickness of the pillbox approaches zero, the integral over S_3 also vanishes. If S_1 is small enough, the integral over S_1 can be approximated by:

$$\oint_S \vec{B} \cdot d\vec{A} =$$

$$= \int_{S_1} \vec{B} \cdot d\vec{A} + \int_{S_2} \vec{B} \cdot d\vec{A} + \int_{S_3} \vec{B} \cdot d\vec{A} = \int_{S_1} \vec{B} \cdot d\vec{A} \cong B_n \cdot A_1 = 0$$

(3.83)

where B_n is the component of the magnetic field normal to the surface and A_1 is the area of S_1. Obviously, this implies that $B_n = 0$. In other words, the normal component of the magnetic field must be zero at the boundary of a perfectly conducting medium.

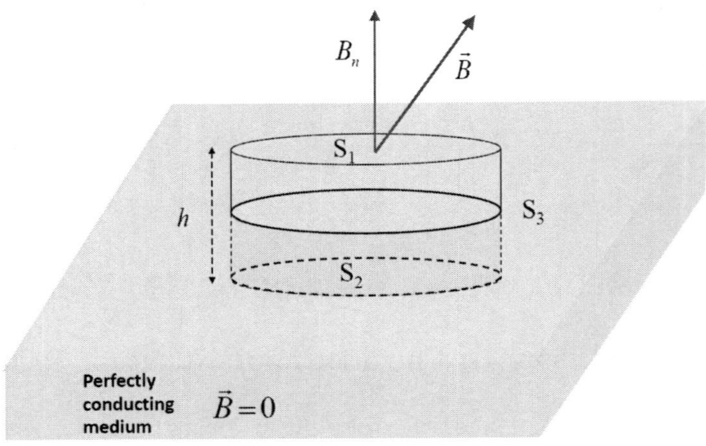

Figure 3.9. Flux through a small 'pillbox' crossed by the surface of a perfectly conducting wall.

3.5.2. Deriving the Fields and the Cutoff Frequencies

Figure 3.10 also shows the magnetic field corresponding to each wave. Again it should be mentioned that the relative orientations of the magnetic field before and after each reflection have not been randomly chosen. The corresponding arrows have been drawn to explicitly illustrate that the normal component of the magnetic field resulting from the incident and reflected waves vanishes at a perfectly conducting wall.

From this figure, we can easily conclude that:

$$B_{1x} = -B_1(\vec{r},t)\sin\phi = -B_0\sin(\vec{k}_1\cdot\vec{r}-\omega t)\sin\phi$$
$$B_{2x} = -B_2(\vec{r},t)\sin\phi = -B_0\sin(\vec{k}_2\cdot\vec{r}-\omega t)\sin\phi$$
$$B_{3x} = B_3(\vec{r},t)\sin\phi = B_0\sin(\vec{k}_3\cdot\vec{r}-\omega t)\sin\phi$$
$$B_{4x} = B_4(\vec{r},t)\sin\phi = B_0\sin(\vec{k}_4\cdot\vec{r}-\omega t)\sin\phi$$

$$(3.84)$$

$$B_{1y} = -B_1(\vec{r},t)\cos\phi = -B_0\sin(\vec{k}_1\cdot\vec{r}-\omega t)\cos\phi$$

$$B_{2y} = B_2(\vec{r},t)\cos\phi = B_0\sin(\vec{k}_2\cdot\vec{r}-\omega t)\cos\phi$$

$$B_{3y} = B_3(\vec{r},t)\cos\phi = B_0\sin(\vec{k}_3\cdot\vec{r}-\omega t)\cos\phi \tag{3.85}$$

$$B_{4y} = -B_4(\vec{r},t)\cos\phi = -B_0\sin(\vec{k}_4\cdot\vec{r}-\omega t)\cos\phi$$

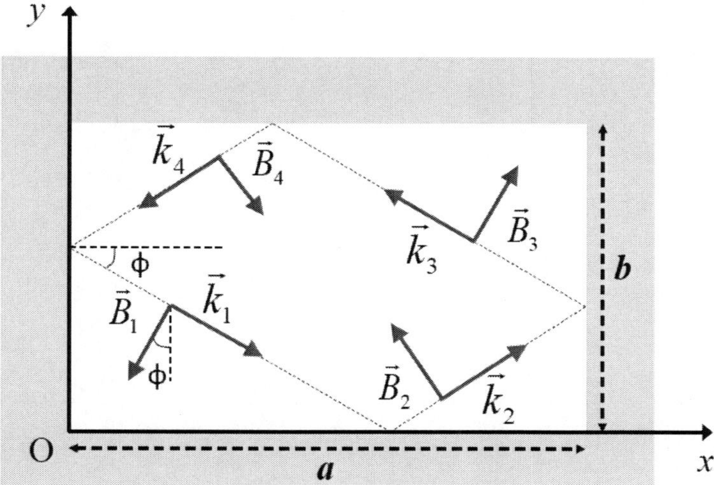

Figure 3.10. Projection of the successive reflections in a rectangular waveguide onto the xy-plane. The wave resulting from the superposition of different waves propagates along the $+z$-direction (not shown). The magnetic field has no component in the direction of propagation (TM mode). However, it should be stressed that the arrows representing wave vectors are only projections and have z-component.

The x- and y-components of the magnetic field can be obtained from the addition of these expressions following a procedure similar to that employed in TE modes (with the help of trigonometric identities):

$$B_x(\vec{r},t) = -4B_0\frac{k_y}{\sqrt{k_x^2+k_y^2}}\sin(k_x x)\cos(k_y y)\cos(k_z z-\omega t)$$

$$\tag{3.86}$$

$$B_y(\vec{r},t) = +4B_0\frac{k_x}{\sqrt{k_x^2+k_y^2}}\cos(k_x x)\sin(k_y y)\cos(k_z z-\omega t)$$

The reader can easily check that:

$$B_x(0, y, z, t) = 0$$
$$B_y(x, 0, z, t) = 0 \tag{3.87}$$

These are the boundary conditions for two of the four walls ($x = 0$ and $y = 0$). The restrictions on the other walls are:

$$B_x(a, y, z, t) = 0$$
$$B_y(x, b, z, t) = 0 \tag{3.88}$$

They can be fulfilled if:

$$k_x = n\pi/a \qquad n = 0,1,2,3,...$$
$$k_y = m\pi/b \qquad m = 0,1,2,3,... \tag{3.89}$$

As can be seen, these equations are identical to those obtained for TE modes. Consequently, the expression for cutoff frequencies remains unaltered.

The Cartesian components of the electric field can be derived from Ampère-Maxwell's law for empty space:

$$\vec{\nabla} \times \vec{B} = \mu_0 \varepsilon_0 \frac{\partial \vec{E}}{\partial t} \tag{3.90}$$

Computing the curl of the magnetic field and solving for the electric field gives:

$$E_x(\vec{r}, t) = \frac{4B_0 c^2}{\omega} \frac{k_x k_z}{\sqrt{k_x^2 + k_y^2}} \cos(k_x x) \sin(k_y y) \cos(k_z z - \omega t)$$

$$E_y(\vec{r}, t) = \frac{4B_0 c^2}{\omega} \frac{k_y k_z}{\sqrt{k_x^2 + k_y^2}} \sin(k_x x) \cos(k_y y) \cos(k_z z - \omega t) \tag{3.91}$$

$$E_z(\vec{r}, t) = \frac{4B_0 c^2}{\omega} \sqrt{k_x^2 + k_y^2} \sin(k_x x) \sin(k_y y) \sin(k_z z - \omega t)$$

All these expression can be written as a function of the amplitude of the z-component, which is given by:

$$E_{z0} \equiv 4B_0 c^2 \sqrt{k_x^2 + k_y^2} / \omega \qquad (3.92)$$

Applying this equation and identifying k_z and k_g gives:

$$B_x(\vec{r},t) = -E_{z0} \frac{\omega}{c^2} \frac{k_y}{k_x^2 + k_y^2} \sin(k_x x)\cos(k_y y)\cos(k_g z - \omega t)$$

$$B_y(\vec{r},t) = E_{z0} \frac{\omega}{c^2} \frac{k_x}{k_x^2 + k_y^2} \cos(k_x x)\sin(k_y y)\cos(k_g z - \omega t)$$

$$E_x(\vec{r},t) = E_{z0} \frac{k_x k_g}{k_x^2 + k_y^2} \cos(k_x x)\sin(k_y y)\cos(k_g z - \omega t) \qquad (3.93)$$

$$E_y(\vec{r},t) = E_{z0} \frac{k_y k_g}{k_x^2 + k_y^2} \sin(k_x x)\cos(k_y y)\cos(k_g z - \omega t)$$

$$E_z(\vec{r},t) = E_{z0} \sin(k_x x)\sin(k_y y)\sin(k_g z - \omega t)$$

Again we should bear in mind that k_x and k_y are quantized, and their values are given by Equation 3.72. Figure 3.11 shows the magnetic field lines for TM_{11} and TM_{21} modes (at a given time).

These patterns obey the rules of thumb for magnetic field lines discussed previously and corroborated for the parallel-plate guide:

i)   Magnetic field lines have no start or end. They form closed loops (this is the case) or extend to infinity.

ii) Near the guide walls, these lines are tangent to the perfectly conducting surface. This property can be explain recalling that, according to the boundary condition deduced in section 3.5.1, the normal component of the magnetic field must be zero at the boundary of a perfectly conducting medium.

In addition, note how the spatial pattern changes when m varies from 1 to 2. The picture for $m = 1$ only contains one group of closed loops circulating in the same direction whereas the pattern for $m = 2$ consists in two groups of continuous loops circulating in opposite directions.

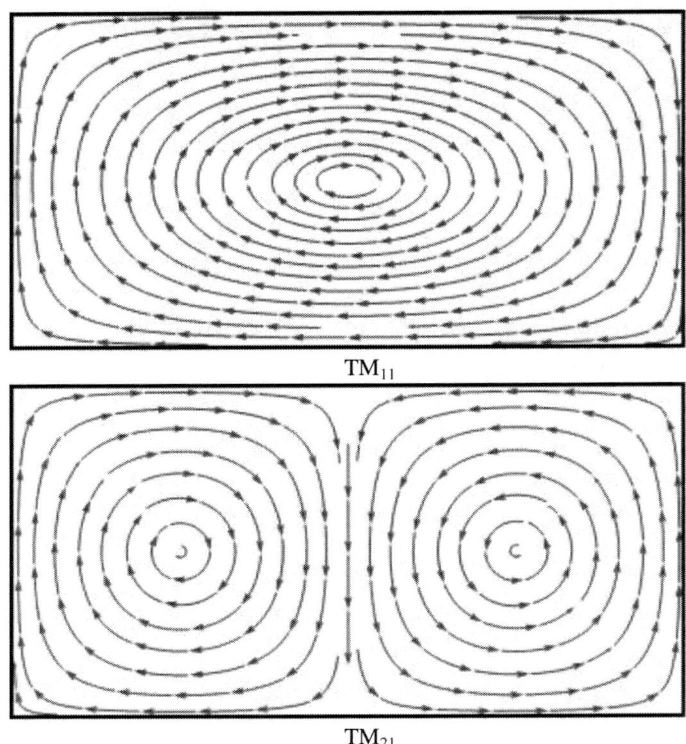

Figure 3.11. Magnetic field lines for TM$_{11}$ and TM$_{21}$ modes for a rectangular waveguide with $a=2b$. Figure generated by Mathematica 8, a registered trademark of Wolfram Research, Inc.

3.5.3. Do TEM Modes Propagate Along a Rectangular Waveguide?

In section 3.3.3 we concluded that TM$_0$ modes can propagate in a parallel-plate waveguide, being characterized by transverse electric and magnetic fields. Thus the following question arises: Can TM$_{00}$ modes propagate in a rectangular waveguide? Equations 3.93 provide the answer setting k_x and k_y zero, which leads to non-existent electric and magnetic fields. Consequently, TEM modes do propagate along a parallel-plate waveguide but not along a rectangular waveguide.

In general, TEM modes are not possible in hollow waveguides confined by a *single* conductor but can propagate along guides that confine between two conductors, such as a coaxial cable.

3.6. SUMMARY

As can be concluded from this chapter, there are several differences between guided and free electromagnetic waves, which are summarized below:

1. Guided waves propagate along the guiding direction with a characteristic wavenumber, which generally differs from the wavenumber in free space.
2. In general, this characteristic wavenumber and frequency are related through a nonlinear relationship known as dispersion relation.
3. Each propagation mode has a cutoff frequency. A wave will propagate along the guide if its frequency is greater than the cutoff frequency.
4. Cutoff frequencies and cutoff wavenumbers of TE and TM modes are quantized.
5. In a waveguide, the electromagnetic signal and energy cannot propagate more rapidly than light. Their speed is known as group velocity. On the contrary, the so-called phase velocity can be greater than the speed of light. These two velocities are identical in free space or TEM modes.

APPENDIX 3.1: WAVE EQUATION IN SEVERAL DIMENSIONS

Dealing with an only plane wave, it is always possible to choose a coordinate system with one of its axes (say the x-axis) in the direction of propagation. In this way, the wave equation can be written as a function of an only space variable, $\xi = f(\omega t - kx)$, where ξ stands for the traveling disturbance associated to the wave (for example, a varying electric field). However, this is not possible dealing with two or more waves propagating in different directions.

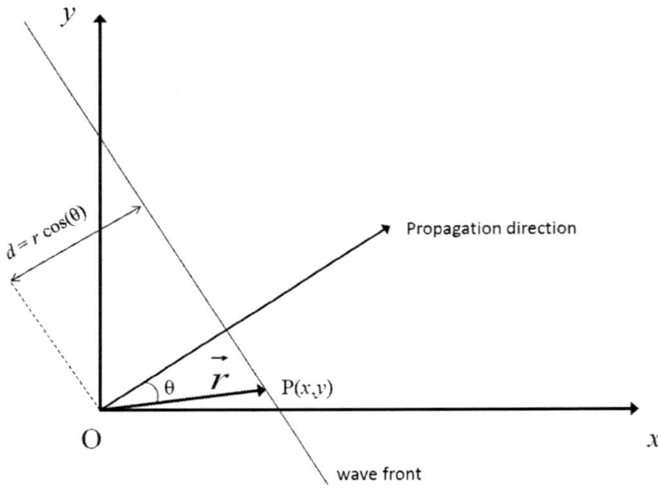

Figure 3.12. Wavefront of a two-dimensional wave. The position vector of a point of this wavefront forms an angle θ with the direction of propagation.

In this appendix we derive an equation for a two-dimensional wave whose direction of propagation is arbitrary. First, let us note that all the points in a plane perpendicular to the direction of propagation are in the same state of disturbance, ξ, since such a plane is a wavefront. Thus x in $\xi = f(\omega t - kx)$ should be replaced by another space variable characterizing the plane. This variable is the distance from the origin, symbolized as d. Thus we can write $\xi = f(\omega t - kd)$. Given a point P whose Cartesian coordinates are (x, y) and a wave whose direction of propagation is described by the unit vector $\hat{u}$, how can we compute d? It can be easily inferred from Figure 3.12 that $d = r\cos\theta$, where r is the magnitude of the position vector of P and θ the angle formed by the direction of propagation with the position vector. d can also be expressed as the scalar product $d = \vec{r} \cdot \hat{u}$, where $\vec{r} = x\hat{i} + y\hat{j}$. Thus we finally end up with:

$$\xi = f(\omega t - kr\cos\theta) = f(\omega t - k\hat{u} \cdot \vec{r}) \tag{3.94}$$

This result can be easily generalized for a three-dimensional wave.

HISTORICAL NOTE: THE BEGINNING OF TELECOMMUNICATIONS AND THE DISCOVERY OF THE LARGEST WAVEGUIDE IN THE WORLD

At the end of 19th century, there was evidence of the existence of electromagnetic waves and the possibility of generating them artificially immediately suggested the idea of using these waves with the aim of transmitting information at a distance. It was beginning the age of telecommunications, whose initial development was favored by the discovery of the ionosphere, a region of the upper atmosphere that together with the earth constitutes the largest waveguide in the world.

Further details lead us to Guglielmo Marconi's life and work. This Italian inventor came from a well-off family and had been educated in private schools. He studied physics with famous professors but did not belong to any university. In 1894, at the age of 20, he found a paper by Hertz on generation and detection of the electromagnetic waves predicted by Maxwell. Marconi was able to build some of the equipment described by Hertz on his own and even went further adding a coherer. This device consists of a tube with slightly cramped metal filings, so the resistance of the coherer is high under usual conditions. However, the initial resistance of the filings reduces when it is exposed to an electromagnetic wave, allowing an electric current to flow through it. In 1895 Marconi tested his new instrumentation and succeeding in transmitting a signal from his house to the garden and, later, a mile away. In 1896 he went to England and transmitted a signal over a distance of about nine miles. In 1897, he transmitted signals to a ship twelve miles away. One year later, the signals covered distances of about eighteen miles. Marconi then claimed that electromagnetic signals could travel distances of about 150 km but many scientists remained skeptical and argued that signals propagated in straight line and could not bend following the curvature of earth.

Marconi went on performing more experiments and slowly increasing the range of transmissions. He also observed that if the transmitter was earthed and connected to a long rod, waves seemed to be guided around the earth surface. In this way, he transmitted signals over really long distances. In 1898 signals crossed the English Channel and in 1901 he performed the first transatlantic transmission, from Poldhu, Cornwall (England) to St John's, Newfoundland (Canada). Due to this success in long-distance transmissions, some scientists tried to find out why electromagnetic wave could find a way around the earth beyond the horizon. Arthur Edwin Kennelly and Oliver

Heaviside independently proposed that there was a layer that contained charged particles and behaved as a conducting medium. This layer was discovered some years later and nowadays is known as the ionosphere. The space between the ground and the boundary of the ionosphere behaves as a large waveguide, as illustrated in Figure 3.13. The signals transmitted by a station A cannot reach station E through a straight line because they would be blocked by the earth at B'. However, the signals emitted from A can be reflected at B and after successive reflections (C and D) bouncing up and down, they reach station E. This is the mechanism discussed in this chapter operating in the largest waveguide. Around 1925, scientists began to unravel the structure and properties of the ionosphere. This layer of the atmosphere is located from about 80 to 500 km altitude and is ionized due to solar radiation.

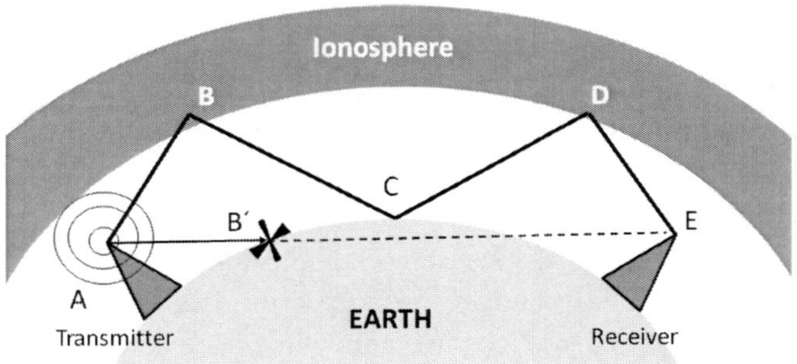

Figure 3.13. The signals transmitted by a station A cannot reach station E through a straight line because they would be blocked by the earth at B'. However, the signals emitted from A can be reflected at B and after successive reflections (C and D) bouncing up and down, they reach station E.

SOLVED PROBLEMS

Problem 3.1. Consider an electromagnetic wave propagating through the parallel-plate waveguide described in section 3.2. Prove that for TE modes the magnetic field turns out to be:

$$\vec{B}(x, y, t) = 2E_0(k_g / \omega)\sin(k_c x)\sin(\omega t - k_g y + \phi)\hat{i}$$

$$- 2E_0(k_c / \omega)\cos(k_c x)\cos(\omega t - k_g y + \phi)\hat{j}$$

Then, check that this field verifies Gauss's law for magnetism and Ampère-Maxwell law.

Solution. Our starting point will be the electric field for TE modes:

$$E_z(x,y,t) = 2E_0 \sin(k_c x) \sin(\omega t - k_g y + \phi)$$

As mentioned in section 3.2, the magnetic field can be computed from Faraday's law:

$$\vec{\nabla} \times \vec{E} = \hat{i}\,\frac{\partial E_z}{\partial y} - \hat{j}\,\frac{\partial E_z}{\partial x} = -\frac{\partial \vec{B}}{\partial t}$$

Solving for the magnetic field leads to:

$$\vec{B} = \int \left(-\hat{i}\,\frac{\partial E_z}{\partial y} + \hat{j}\,\frac{\partial E_z}{\partial x} \right) dt$$

After taking partial derivatives we obtain:

$$\vec{B} = \hat{i} \int \left(2E_0 k_g \sin(k_c x) \cos(\omega t - k_g y + \phi) \right) dt$$
$$+ \hat{j} \int \left(2E_0 k_c \cos(k_c x) \sin(\omega t - k_g y + \phi) \right) dt$$

Integrating with respect to time gives:

$$\vec{B} = 2E_0 (k_g / \omega) \sin(k_c x) \sin(\omega t - k_g y + \phi)\hat{i}$$
$$- 2E_0 (k_c / \omega) \cos(k_c x) \cos(\omega t - k_g y + \phi)\hat{j}$$

Our function depends on x, y and t but the integration was performed only with respect to t. Thus we may consider the addition of a vector function depending on x and y. However, this function does not depend on t and cannot be properly considered a traveling wave solution. For our purposes, it can be dropped.

It can be easily shown that this field satisfies Gauss's law for magnetism. Taking partial derivatives and adding yields:

$$\vec{\nabla} \cdot \vec{B} = \frac{\partial B_x}{\partial x} + \frac{\partial B_y}{\partial y} + \frac{\partial B_z}{\partial z} =$$

$$= 2E_0(k_g k_c / \omega)\cos(k_c x)\sin(\omega t - k_g y + \phi)$$

$$- 2E_0(k_c k_g / \omega)\cos(k_c x)\sin(\omega t - k_g y + \phi) = 0$$

Finally, let us verify Ampère-Maxwell's law. The curl of this field is:

$$\vec{\nabla} \times \vec{B} = \begin{vmatrix} \hat{i} & \hat{j} & \hat{k} \\ \frac{\partial}{\partial x} & \frac{\partial}{\partial y} & 0 \\ B_x & B_y & 0 \end{vmatrix} = \left(\frac{\partial B_y}{\partial x} - \frac{\partial B_x}{\partial y} \right)\hat{k} =$$

$$+ 2E_0(k_c^2 / \omega)\sin(k_c x)\cos(\omega t - k_g y + \phi)\hat{k}$$

$$+ 2E_0(k_g^2 / \omega)\sin(k_c x)\cos(\omega t - k_g y + \phi)\hat{k}$$

Factoring and recalling that $k_g^2 + k_c^2 = k^2$ leads to:

$$\vec{\nabla} \times \vec{B} = 2E_0(k^2 / \omega)\sin(k_c x)\cos(\omega t - k_g y + \phi)\hat{k}$$

The other side of Ampère-Maxwell is:

$$\mu_0 \varepsilon_0 \frac{\partial \vec{E}}{\partial t} = \mu_0 \varepsilon_0 \omega \cdot 2E_0 \sin(k_c x)\cos(\omega t - k_g y + \phi)\hat{k} =$$

$$= \frac{\omega}{c^2} 2E_0 \sin(k_c x)\cos(\omega t - k_g y + \phi)\hat{k} =$$

$$= 2E_0(k^2 / \omega)\sin(k_c x)\cos(\omega t - k_g y + \phi)\hat{k}$$

As can be seen, both sides are identical:

$$\vec{\nabla} \times \vec{B} = \mu_0 \varepsilon_0 \frac{\partial \vec{E}}{\partial t}$$

> **Problem 3.2.** Consider again an electromagnetic wave propagating through the parallel-plate waveguide described in sections 3.2 and 3.3. Find the electric field for TM modes.

Solution. The procedure for solving this problem is similar to that employed in Problem 3.1 and begins with the magnetic field for TM modes:

$$B_z(x, y, t) = 2B_0 \cos(k_c x) \sin(\omega t - k_g y + \phi)$$

In this case, however, the electric field will be calculated from Ampère-Maxwell law. First, let us work out the curl of the magnetic field:

$$\vec{\nabla} \times \vec{B} = \hat{i}\,\frac{\partial B_z}{\partial y} - \hat{j}\,\frac{\partial B_z}{\partial x} = \mu_0 \varepsilon_0 \frac{\partial \vec{E}}{\partial t}$$

Solving for the electric field gives:

$$\vec{E} = c^2 \int \left(\hat{i}\,\frac{\partial B_z}{\partial y} - \hat{j}\,\frac{\partial B_z}{\partial x} \right) dt$$

Taking partial derivatives:

$$\vec{E} = c^2 \int \hat{i}(-2B_0 k_g \cos(k_c x)\cos(\omega t - k_g y + \phi)) dt$$
$$+ c^2 \int \vec{j}(2B_0 k_c \sin(k_c x)\sin(\omega t - k_g y + \phi)) dt$$

And integrating with respect to time gives:

$$\vec{E} = -2B_0(k_g / \omega)c^2 \cos(k_c x)\sin(\omega t - k_g y + \phi)\hat{i}$$
$$- 2B_0(k_c / \omega)c^2 \sin(k_c x)\cos(\omega t - k_g y + \phi)\vec{j}$$

Note that the integration constant was dropped because it cannot be considered a traveling wave solution.

Problem 3.3. Consider the air-filled rectangular waveguide known as WR 90, whose dimensions are 2.286 and 1.016 cm. This guide is useful for military radars working in the X band. It is desired to design this guide so that: a) only the dominant mode is transmitted; b) the operating frequency must lie between 25% above the lowest cutoff frequency and 90% of the next cutoff frequency. Determine the lower and upper limits of the operating frequency.

Solution. First, let us compute the lowest frequencies of the modes of this guide, which are required to determine the operating bandwith. These frequencies can be computed through the expression:

$$f_c(TE_{nm}) = f_c(TM_{nm}) = c\sqrt{\left(\frac{n}{2a}\right)^2 + \left(\frac{m}{2b}\right)^2}$$

In this case, $a = 2.286$ cm and $b = 1.016$ cm:

$$f_c(TE_{10}) = f_c(TM_{10}) = 3\cdot10^8\sqrt{\left(\frac{1}{2\cdot2.286\cdot10^{-2}}\right)^2} = 6.56\cdot10^9\,Hz = 6.56\,GHz$$

$$f_c(TE_{20}) = f_c(TM_{20}) = 3\cdot10^8\sqrt{\left(\frac{2}{2\cdot2.286\cdot10^{-2}}\right)^2} = 13.1\cdot10^9\,Hz = 13.1\,GHz$$

$$f_c(TE_{01}) = f_c(TM_{01}) = 3\cdot10^8\sqrt{\left(\frac{1}{2\cdot1.016\cdot10^{-2}}\right)^2} = 14.7\cdot10^9\,Hz = 14.7\,GHz$$

$$f_c(TE_{02}) = f_c(TM_{02}) = 3\cdot10^8\sqrt{\left(\frac{2}{2\cdot1.016\cdot10^{-2}}\right)^2} = 29.5\cdot10^9\,Hz = 29.5\,GHz$$

Note that these frequencies have been arranged in increasing order. The operating bandwidth is determined by the two lowest frequencies:

$$f \in [6.56\,GHz, 13.1\,GHz]$$

But according to the wording, permitted frequencies lie in a more restricted range:

$$f_{per} \in [1.25 \cdot 6.56\,GHz, 0.90 \cdot 13.1\,GHz] = [8.2\,GHz, 11.8\,GHz]$$

> **Problem 3.4.** Consider again the air-filled rectangular waveguide of the previous problem. If this guide operates at 10.4 GHz, calculate the maximum power that can be transmitted without causing dielectric breakdown of air. The dielectric strength of air is 3 MV/m.

As required in the previous problem, only the dominant mode (TE_{10}) is transmitted. In section 3.4, the power transmitted in the TE_{10} mode was calculated, turning out to be:

$$P = \frac{E_{0y}^2 k_g ba}{4\mu_0 \omega}$$

In our case, the maximum amplitude of the electric field is 3 MV/m (to avoid dielectric breakdown of air), $\omega = 6.53 \cdot 10^{10}$ rad/s, $k = 217.8$ m^{-1} and

$$k_g = \sqrt{k^2 - \left(\frac{\pi}{a}\right)^2} = 169.0\,m^{-1}$$

Inserting these values into the expression that provides the transmitted power, we obtain $1 \cdot 10^6$ W (in round numbers).

> **Problem 3.5.** Consider and air-filled rectangular waveguide with dimensions $a = 2.5$ cm and $b = 1.0$ cm. The operating frequency cannot be greater than 8.5 GHz. Which TE and TM modes can it transmit?

Solution. Recall that the cutoff frequencies of TE and TM modes are given by:

$$f_c(TE_{nm}) = f_c(TM_{nm}) = c\sqrt{\left(\frac{n}{2a}\right)^2 + \left(\frac{m}{2b}\right)^2}$$

In our case, $a = 2.5$ cm and $b = 1.0$ cm. Let us compute the cutoff frequencies corresponding to the lowest modes:

$$f_c(TE_{10}) = f_c(TM_{10}) = 3 \cdot 10^8 \sqrt{\left(\frac{1}{2 \cdot 0.025}\right)^2} = 6 \cdot 10^9 \, Hz = 6 \, GHz$$

$$f_c(TE_{01}) = f_c(TM_{01}) = 3 \cdot 10^8 \sqrt{\left(\frac{1}{2 \cdot 0.01}\right)^2} = 15 \cdot 10^9 \, Hz = 15 \, GHz$$

$$f_c(TE_{20}) = f_c(TM_{20}) = 3 \cdot 10^8 \sqrt{\left(\frac{2}{2 \cdot 0.025}\right)^2} = 12 \cdot 10^9 \, Hz = 12 \, GHz$$

As you can see, only TE_{10} and TM_{10} modes can propagate through this guide.

Problem 3.6. a) Determine the dimensions of an air-filled waveguide for which the cutoff frequencies corresponding to the TM_{11} y TE_{30} modes are 12 GHz (in both cases). b) If the operating frequency is 7 GHz, find out if the dominant mode can propagate.

Solution. a) First, we will determine the dimension of the guide from the cutoff frequencies of the TM_{11} y TE_{30} modes recalling that:

$$f_c(TE_{nm}) = f_c(TM_{nm}) = c \sqrt{\left(\frac{n}{2a}\right)^2 + \left(\frac{m}{2b}\right)^2}$$

One of the dimensions, a, can be straightforwardly computed from $f_c(TE_{30}) = 12 \, GHz$:

$$f_c(TE_{30}) = 12 \cdot 10^9 = 3 \cdot 10^8 \cdot \sqrt{\left(\frac{3}{2a}\right)^2}$$

Thus:

$$\frac{3}{2a} = \frac{12 \cdot 10^9}{3 \cdot 10^8} \quad \Rightarrow \quad a = 0.0375 \, m = 3.75 \, cm$$

The other dimension can be determined applying that $f_c(TM_{11}) = 12 \, GHz$:

$$f_c(TM_{11}) = 12 \cdot 10^9 = 3 \cdot 10^8 \sqrt{\left(\frac{1}{2 \cdot 0.0375}\right)^2 + \left(\frac{1}{2b}\right)^2}$$

Solving for b gives:

$$b = 0.01326 \, m = 1.326 \, cm$$

b) Now, let us determine the cutoff frequencies of the lowest modes of this waveguide, which allows us to know the operating bandwidth:

$$f_c(TE_{10}) = f_c(TM_{10}) = 3 \cdot 10^8 \sqrt{\left(\frac{1}{2 \cdot 0.0375}\right)^2} = 4 \cdot 10^9 \, Hz = 4 \, GHz$$

$$f_c(TE_{20}) = f_c(TM_{20}) = 3 \cdot 10^8 \sqrt{\left(\frac{2}{2 \cdot 0.0375}\right)^2} = 8 \cdot 10^9 \, Hz = 8 \, GHz$$

$$f_c(TE_{01}) = f_c(TM_{01}) = 3 \cdot 10^8 \sqrt{\left(\frac{1}{2 \cdot 0.01326}\right)^2} = 11.3 \cdot 10^9 \, Hz = 11.3 \, GHz$$

Again, the modes have been arranged in increasing order. As you can see, $f_c(TE_{10}) \le 7 \, GHz \le f_c(TE_{20})$. Thus the dominant mode can propagate operating at 7 GHz.

> **Problem 3.7.** Find out why waveguides are not used for VHF. A typical frequency belonging to this band is 45 MHz.

The answer to this question can be found imaging a square air-filled wave guide ($a = b$). Let us estimate a representative a-value assuming that the cutoff frequency for TE_{10} is 45 MHz. Accordingly:

$$f_c(TE_{10}) = 45 \cdot 10^6 = c\sqrt{\left(\frac{1}{2a}\right)^2} = 3 \cdot 10^8 \cdot \sqrt{\left(\frac{1}{2a}\right)^2} = 3 \cdot 10^8 \frac{1}{2a}$$

Thus:

$$a = \frac{3 \cdot 10^8}{2 \cdot 45 \cdot 10^6} = 3.33\,m$$

In the view of this result, we can conclude that this guide would have hardly manageable dimensions.

> **Problem 3.8.** We want to design an air-filled rectangular waveguide that must operate between 7.5 and 11 GHz. In addition, the guide must meet the following specifications: a) there must be only one mode of propagation; b) the lower limit of frequencies has to be 10% above the lowest cutoff frequency; c) the upper limit of operating frequencies has to be 90% of the next cutoff frequency. Find out the dimensions of the waveguide.

Solution. The two lowest cutoff frequencies, which we will refer to as f_{c1} and f_{c2}, must satisfy that:

$$7.5\,GHz = 1.1 f_{c1}$$
$$11.0\,GHz = 0.9 f_{c2}$$

From these conditions, f_{c1} and f_{c2}, turn out to be 6.82 GHz and 12.22 GHz, respectively. Thus:

$$f_c(TE_{10}) = 6.82 \cdot 10^9 = c\sqrt{\left(\frac{1}{2a}\right)^2} = \frac{c}{2a} = \frac{3 \cdot 10^8}{2a}$$

$$\Rightarrow \quad a = 0.0220\,m = 2.20\,cm$$

$$f_c(TE_{01}) = 12.22 \cdot 10^9 = c\sqrt{\left(\frac{1}{2b}\right)^2} = \frac{c}{2b} = \frac{3 \cdot 10^8}{2b}$$

$$\Rightarrow \quad b = 0.0123\,m = 1.23\,cm$$

Let us check that the cutoff frequency corresponding to the TE_{20} mode is greater than the upper limit:

$$f_c(TE_{20}) = c\sqrt{\left(\frac{2}{2a}\right)^2} = \frac{c}{a} = \frac{3 \cdot 10^8}{0.0220} = 13.6\,GHz$$

Problem 3.9. Prove that if the operating bandwidth of an air-filled rectangular waveguide is given by $f \in \left[f_c(TE_{10}), f_c(TE_{01})\right]$ then $b > a/2$.

Solution. If the bandwidth is determined by the cutoff frequencies corresponding to TE_{10} y TE_{01} modes, this means that:

$$f_c(TE_{20}) > f_c(TE_{01})$$

Thus:

$$c\sqrt{\left(\frac{2}{2a}\right)^2} > c\sqrt{\left(\frac{1}{2b}\right)^2} \quad \Rightarrow \quad \frac{1}{a} > \frac{1}{2b} \quad \Rightarrow \quad b > \frac{a}{2}$$

Problem 3.10. Consider TE modes in the rectangular waveguide analyzed in section 3.4. Prove explicitly that the y-component of the electric field is given by:

$$E_y(\vec{r},t) = 4E_0 \frac{k_x}{\sqrt{k_x^2 + k_y^2}} \sin(k_x x)\cos(k_y y)\cos(k_z z - \omega t)$$

Solution. Our starting point is Equation 3.68:

$$E_{1y} = E_1(\vec{r},t)\cos\phi = E_0 \sin(\vec{k}_1 \cdot \vec{r} - \omega t)\cos\phi$$

$$E_{2y} = E_2(\vec{r},t)\cos\phi = E_0 \sin(\vec{k}_2 \cdot \vec{r} - \omega t)\cos\phi$$

$$E_{3y} = -E_3(\vec{r},t)\cos\phi = -E_0 \sin(\vec{k}_3 \cdot \vec{r} - \omega t)\cos\phi$$

$$E_{4y} = -E_4(\vec{r},t)\cos\phi = -E_0 \sin(\vec{k}_4 \cdot \vec{r} - \omega t)\cos\phi$$

After adding and factoring, the resulting y-component is:

$$E_y(\vec{r},t) = E_{1y} + E_{2y} + E_{3y} + E_{4y} =$$
$$E_0 \cos\phi[\sin(k_x x - k_y y + k_z z - \omega t) + \sin(k_x x + k_y y + k_z z - \omega t)$$
$$- \sin(-k_x x + k_y y + k_z z - \omega t) - \sin(-k_x x - k_y y + k_z z - \omega t)]$$

This expression can be simplified applying the trigonometric identity for the addition of two sine functions to the two couples of sine functions:

$$E_y(\vec{r},t) = 2E_0 \cos\phi[\sin(k_x x + k_z z - \omega t)\cos(k_y y)$$
$$- \sin(-k_x x + k_z z - \omega t)\cos(k_y y)] =$$
$$2E_0 \cos\phi[\sin(k_x x + k_z z - \omega t)\cos(k_y y)$$
$$+ \sin(k_x x - k_z z + \omega t)\cos(k_y y)] =$$

After factoring $\cos(k_y y)$, the same identity can be applied again, which yields:

$$E_y(\vec{r},t) = 4E_0 \cos\phi \sin(k_x x)\cos(k_y y)\cos(k_z z - \omega t)$$

Expressing $\cos\phi$ in terms of the Cartesian components of the wave vector leads to the result that we want to prove.

Problem 3.11. Consider again TE modes in the rectangular waveguide analyzed in section 3.4. Prove explicitly that the Cartesian components of the magnetic field are given by Equation 3.76.

Solution. Here we will apply Faraday's law. First, the curl of the electric field will be computed. This field (for TE modes) is given by (see Equations 3.67 and 3.69):

$$E_x(\vec{r},t) = -4E_0 \frac{k_y}{\sqrt{k_x^2 + k_y^2}}\cos(k_x x)\sin(k_y y)\cos(k_z z - \omega t)$$

$$E_y(\vec{r},t) = 4E_0 \frac{k_x}{\sqrt{k_x^2 + k_y^2}}\sin(k_x x)\cos(k_y y)\cos(k_z z - \omega t)$$

Its curl is calculated expanding the determinant and taking partial derivatives:

$$\vec{\nabla} \times \vec{E} = \begin{vmatrix} \hat{i} & \hat{j} & \hat{k} \\ \frac{\partial}{\partial x} & \frac{\partial}{\partial y} & \frac{\partial}{\partial z} \\ E_x & E_y & 0 \end{vmatrix} = -\frac{\partial E_y}{\partial z}\hat{i} + \frac{\partial E_x}{\partial z}\hat{j} + \left(\frac{\partial E_y}{\partial x} - \frac{\partial E_x}{\partial y}\right)\hat{k} =$$

$$= 4E_0 \frac{k_x k_z}{\sqrt{k_x^2 + k_y^2}} \sin(k_x x)\cos(k_y y)\sin(k_z z - \omega t)\hat{i}$$

$$+ 4E_0 \frac{k_y k_z}{\sqrt{k_x^2 + k_y^2}} \cos(k_x x)\sin(k_y y)\sin(k_z z - \omega t)\hat{j}$$

$$+ 4E_0 \frac{k_x^2}{\sqrt{k_x^2 + k_y^2}} \cos(k_x x)\cos(k_y y)\cos(k_z z - \omega t)\hat{k}$$

$$+ 4E_0 \frac{k_y^2}{\sqrt{k_x^2 + k_y^2}} \cos(k_x x)\cos(k_y y)\cos(k_z z - \omega t)\hat{k}$$

After factoring, the z-component can be simplified, which yields:

$$\vec{\nabla} \times \vec{E} = 4E_0 \frac{k_x k_z}{\sqrt{k_x^2 + k_y^2}} \sin(k_x x)\cos(k_y y)\sin(k_z z - \omega t)\hat{i}$$

$$+ 4E_0 \frac{k_y k_z}{\sqrt{k_x^2 + k_y^2}} \cos(k_x x)\sin(k_y y)\sin(k_z z - \omega t)\hat{j}$$

$$+ 4E_0 \sqrt{k_x^2 + k_y^2}\, \cos(k_x x)\cos(k_y y)\cos(k_z z - \omega t)\hat{k}$$

Solving for the magnetic field in Faraday's law gives:

$$\vec{B} = -\int \vec{\nabla} \times \vec{E} dt$$

Taking into account that:

$$-\int \sin(k_z z - \omega t)dt = -\frac{1}{\omega}\cos(k_z z - \omega t)$$

$$-\int \cos(k_z z - \omega t)dt = \frac{1}{\omega}\sin(k_z z - \omega t)$$

We end up with:

$$\vec{B} = -4E_0 \frac{k_x k_z}{\omega\sqrt{k_x^2 + k_y^2}} \sin(k_x x)\cos(k_y y)\cos(k_z z - \omega t)\hat{i}$$

$$-4E_0 \frac{k_y k_z}{\omega\sqrt{k_x^2 + k_y^2}} \cos(k_x x)\sin(k_y y)\cos(k_z z - \omega t)\hat{j}$$

$$+4E_0 \frac{\sqrt{k_x^2 + k_y^2}}{\omega} \cos(k_x x)\cos(k_y y)\sin(k_z z - \omega t)\hat{k}$$

FURTHER READING

A mentioned at the preface, this book tries to bridge the gap between elementary and advanced treatments on Maxwell's equations and propagation of electromagnetic waves. You can find the integral form of Maxwell's equations in many excellent textbooks of introduction to physics for undergraduates. Here we only mention a couple of them. This choice is just a matter of opinion:

- Tipler, P.A., Mosca, G., (2007), *Physics for Scientists and Engineers* (sixth edition), New York, United States of America, W. H. Freeman and Company.
- Young, H.D., Freedman, R.A., (2004), *University Physics Volume 2*(eleventh edition), San Francisco, United States of America, Addison Wesley.

Dealing with Maxwell' equations, we must recommend the book:

- Fleisch, D., (2008), *A Student's Guide to Maxwell's Equations*, New York, United States of America, Cambridge University Press.

This text is recommendable because it includes a comprehensive description of the notation used in Maxwell's equation as well as comments on their physical meaning. Both integral and differential forms are addressed. In this sense, the text is self-contained. However, the differential form is not derived from the integral formalism. We should also mention that i) the wave equation is derived from advanced vector calculus identities, as usual; ii) guided waves are not treated.

On the other hand, there also are outstanding intermediate-level textbooks on advanced electromagnetism that go beyond the propagation of plane waves in empty space. Here we restrict ourselves to recommend one that includes a chapter devoted to vector operators and many examples throughout the book:

- Cheng, D. K., (1993), *Fundamentals of Engineering Electromagnetics*, United States of America, Addison-Wesley.